HAUNEBU: THE SECRET FILES

The Greatest UFO Secret of All Time

David Hatcher Childress

Adventures Unlimited Press

HAUNEBU: THE SECRET FILES

The Greatest UFO Secret of All Time

Adventures Unlimited Press

Haunebu: The Secret Files

ISBN 978-1-948803-31-1

Published by:
Adventures Unlimited Press
One Adventure Place
Kempton, Illinois 60946 USA
auphq@frontiernet.net

Cover by Terry Lamb

AdventuresUnlimitedPress.com

10 9 8 7 6 5 4 3 2 1

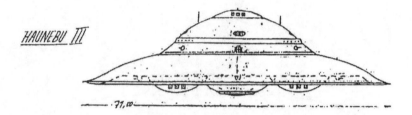

HAUNEBU III

71,00

SCHWERER BEWAFFNETER FLUGKREISEL "HAUNEBU III"

Durchmesser: 71 Meter
Antrieb: Thule-Tachionator 7c plus Schumann-Levitatoren (gepanzert)
Steuerung: Mag-Feld-Impulsor 4a.
Geschwindigkeit: ca. 7000 Kilom. p.Stunde (rechnerisch bis zu 40000)
Reichweite (in Flugdauer): ca. 8 Wochen (bei H-L-Flug 40% mehr)
Bewaffnung: 4 x 11cm KSK in Drehtürmen (3 unten, 1 oben), 10 x 8cm KSK
in Drehringen plus 6 x MK 108, 8 x Jon KSK ferngesteuert
Außenpanzerung: Braischott-Viktalen
Besatzung: 32 Mann (als Transportvar. max. 70 Personen)
Vollsichtfähigkeit: 100 %.
Stilleschwebefähigkeit: 25 Minuten.
Allgemeines Flugvermögen: Wetterunabhängig Tag und Nacht
grundsätzliche Einsatztauglichkeit: Etwa 1945.

Bemerkung: SS-E-IV hält den Hinweis für notwendig, daß in
"Haunebu III" ein großartiges Werk deutscher Technik in entstehen
stehen soll, wegen der allgemeinen Materiallage aber alle
Kräfte auf das schneller verfügbare Haunebu II gesetzt
werden sollten.
Gemeinsam mit dem leichten Flugkreisel "Vril" der Schumann-
Gruppe könnte "Haunebu II" die vom Führer aufgestellten
Forderungen sicherlich erfüllen.

Plans for the Haunebu II.

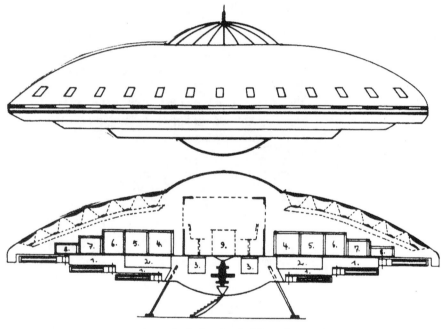

An early design for the Haunebu.

HAUNEBU: THE SECRET FILES

The Greatest UFO Secret of All Time

David Hatcher Childress

HAUNEBU III

Plans for the Haunebu III.

TABLE OF CONTENTS

A Haunebu II on the ground.

A Haunebu II in flight. Note the canon beneath the craft.

Chapter 1

A Saucer Full of Secrets

Dark star crashes, pouring its light into ashes.
Reason tatters, the forces tear loose from the axis.
Searchlight casting for faults in the clouds of delusion.
Shall we go, you and I while we can—
Through the transitive nightfall of diamonds?
–*Dark Star*, Grateful Dead

The mystery of the saucers actually began before WWII, but it was during this war, and shortly afterwards that the phenomena of flying saucers, flying boomerangs, flying cigar-shaped objects and more began to flood the skies of the world. What on earth was going on? Were we being invaded from another planet? According to the press and the military, no one knew for sure.

Were there really UFOs in the skies? In later years, various skeptics and skeptical societies promoted the view that there weren't really unexplained flying machines in the sky, it was just the overactive imaginations of people in the years after WWII. Yet, is this really a good explanation for a large part of the UFO phenomena? Certainly not. Evidence does exist that craft from other planets are visiting the Earth. Evidence also exists that the Germans were designing and building disk-type craft during WWII and perhaps before.

A Document Dump of German Flying Saucers

Since about 1990, researchers in Europe, Australia, America and elsewhere have received documents that show the designs of German flying saucers. A number of curious photos of the craft, in flight and on the ground, have also surfaced. They arrived via a curious document dump involving a German living in London named Ralf Ettl. Ralf Ettl then teamed up with another German named Norbert Jürgen Rathoffer (sometimes spelled Ratthoffer, with a double t, and sometimes with only one f, and sometimes with a hyphen between Jürgen and Rathoffer).

The two released a book, circa 1989, titled *UFO—Das Dritte Reich schlägt zurück?* (*UFO—The Third Reich Strikes Back?*[21]). This book published photos and plans of the Haunebu, Vril and Andromeda craft. They went on to author and publish a second book in 1992 entitled *Das Vril-Projekt: Der Endkampf um die Erde* (*The Vril Project: The Final Battle for the Earth*[22]) That same year they also released a one-hour documentary film in German called *UFO—Geheimnisse des 3 Reichs* (*UFO—Secrets of the Third Reich*). This film was released in Austria in 1992 but saw limited release. It was sold as a videotape and it seems doubtful that it was ever shown on television in Austria, but may have been sold to television stations in other countries.

These Haunebu and Vril documents were published again in 1996 in a German language book called *Die Dunkle Seite Des Mondes* (*The Dark Side of the Moon*)[23] by "Brad Harris" and published by Pandora Books in Germany. This German language book purported to be the translation of an English book, supposedly published a few years earlier.

However, no such English book ever existed and it would seem that "Brad Harris" is just a fake name for some German author, perhaps Norbert Jürgen Rathoffer. The book contained a number of Haunebu and Vril documents and I republished some of these illustrations in the updated version of my book *Man-Made UFOs*.[16] This was a clever way of bringing controversial documents out—in Germany no less—of the Haunebu and Vril craft that had been rumored for decades. Germany has strict laws concerning symbols and speech surrounding the Nazi era and the swastika is banned from public use, including in magazines, books and posters. Photos of the Haunebu or Vril craft do not have swastikas on them allowing them to be viewed in Germany. When they have a symbol on them it is the German Cross. It was later said that the Haunebu and Vril craft had the Black Sun logo on them, instead of the German Cross. Austria does not have the same laws forbidding discussion or images of the Nazis and this is the reason that the film *UFO—Geheimnisse des 3. Reichs* was released in Austria and not Germany.

This movie can be found on the IMDB website where we learn that the English language version, called *UFO Secrets of the Third Reich,* was released in Austria in 1998. The brief description says, "What did the Germans and Hitler know about the universe and UFOs? This documentary unravels some of the mysterious knowledge that was present during the Third Reich." We are told the two writers are Jürgen Rathoffer (screenplay), Ralf Ettl (screenplay). We are also told that the documentary features Ernst Harmannstein, Wolfgang Pampel, Eugen Lardy, and a number of others.

So, clearly Norbert Jürgen Rathoffer and Ralf Ettl are behind the release of most of the Haunebu and Vril documents, but are they trustworthy? Who is behind this document dump? An English version of this documentary, released in Australia in 1996, had a narrator at the beginning announce that the material had come from "the Austrian Branch of the Knights Templar." Some have claimed that the material came from the Austrian branch of the Vril Society.

The American with the most knowledge of these documents is Henry Stevens, a Californian researcher who has been studying the field of German flying disks and other secret technology for decades. Stevens has been in contact with several German researchers on this subject and is the author of a number of books, including *Hitler's Flying Saucers*[34] and *Dark Star*[46].

In private correspondence Stevens told me:

> I was suspicious of the "Ettl Dump" for years. I still am. It is just a little too good. What made me reappraise my position was the late Heiner Gehring who was a straight up, no BS guy. As I mentioned he visited Landig [a former SS officer and author of fictional books on German saucers and secret bases] just before Landig died and was shown Landig's files. Evidently, there was a lot Landig did not tell us in his novels. Heiner came away talking about the Virl and Haunebu saucers as if they were real.

13

Then there is the field propulsion saucer FBI report (the Gut Alt Glossen Report) in my first book [*Hitler's Flying Saucers*] which fits the bill exactly.

I do not know that Heiner believed all of the Tempelhofgesellschaft lore about the Haunebus, the mediums, the fight to other worlds, etc. I rather doubt he believed this and I certainly do not believe this. Nevertheless, he reported some of it, probably based on what he found with Landig.

Heiner was my source as to the Ettl Dump. He explained it to me. In themselves, Ettl and Rathoffer are not credible sources in my eyes. Rathoffer was a security guard or janitor who claimed membership in this secret society. Ettl was a filmmaker and wannabe big time producer. Let me give you an example.

Ettl's production/movie company filed bankruptcy in Germany. This is a very big deal in Germany. But Ettl still wanted to make money on the films belonging to the company and had the master tapes. He sold the right to copy them to a guy in New Zealand or Australia. I found out about this, realized it is probably illegal but my friend at that time Wolf Leithardt, a German living near me, wanted the rights too. They got together and for $500.00 he bought the rights to *UFOs Secrets of the 3rd Reich*— but not the exclusive rights. So Ettl was going around the world selling rights to this film that he made but technically was part of a bankruptcy.

Add to this Ettl and Rathoffer had some sort of interest in a post-production company in Austria, one which could do digital images. We called that company at the time and when the name of the first film or the names of Ettl and Rathoffer were mentioned, people could not run away fast enough.

[Bulgarian-American researcher] Vladimir Terziski was given a couple pictures of a Haunebu in NEGATIVE form. This had great credibility in my mind since I do not recall anyone faking a negative. I think these pictures were in my first book [*Hitler's Flying Saucers*]. Vladimir

14

visited both these guys with that Japanese UFO film crew. Vladimir went to college in Japan and spoke Japanese. But neither Ettl or Rathoffer spoke English so they found a local guy to translate to English for the visit. It was Jan van Helsing the conspiracy writer aka Udo Holy.

About this time I began to turn more and more to the people researching the Jonas Valley as better sources in Germany as well as Friedrich Georg. These guys did not deny the Haunebu thing but like me tried to keep it at an arm's distance. The only one who accepted the Haunebu narrative wholly was Dr. Axel Stoll, a geophysicist.

I still do not know for sure anything about the whole Ettl/Rathoffer/Haunebu thing.

Then in a second email he said:

I forgot two things.

1. Vesco's Kugelblitz (ball lightning) was a reference to the plasma nature of the Haunebu engine. I never made this point but I should have done so.

2. Michael X. Barton's German Saucer Story contained a description of the Haunebu. He listed several, maybe 8 or so, wonder weapons. One he called it the "magnetic bottle" or the flying magnetic bottle. This was some sort of flying craft which sounds very Haunebu-like.

So, Henry Stevens believes that the documents concerning the Haunebu, Vril and Andromeda craft are authentic, though questions remain as to their origins; he doesn't think much of the men who released them. He mentions the field propulsion saucer FBI report, which he calls the Gut Alt Glossen Report, as something that convinced him that the Haunebu saucer had been built and used during the war and after.

The Gut Alt Glossen Report was actually two FBI reports from November 7 and 8, 1957. It is from these reports from the Freedom of Information Act that reports of Haunebu craft in action can be learned. The recently released CIA reports are another. The FBI reports said that a witness saw a craft identical to a Haunebu at the

15

German town of Gut Alt Golssen. Says Stevens:

> The files in question are Gut Alt Golssen file numbers 62-83894-383, 62-838994-384 and 62-83894-385. Their date is 11/7/57 to 11/8/57. They deal with a Polish immigrant, then living in the United States, who reported his wartime experience to the Bureau hoping it might throw some light on UFO sightings seen in Texas at about this time.
>
> The time of the sighting was in 1944, the place was Gut Alt Golssen, approximately 30 miles east of Berlin. The informant, whose name has been deleted, states that while he was a prisoner of war working for the Germans, a flying object arose nearby from behind an enclosure hidden from view by a 50-foot high tarpaulin-type wall. It rose about 500 feet then moved away horizontally. The only noise the object made was a high-pitched whine. The object was described as being 75 to 100 feet in diameter and 14 feet high. It was composed of a dark grey stationary top and bottom sections five to six feet high with a rapidly moving center section producing only a blur and extending the circumference of the vehicle. Notably, the engine of their farm tractor stalled during this event and the SS guards told the driver not to attempt a restart until the whine could no longer be heard.
>
> Because of what I believe is their importance, these files have been reproduced here… One of the most compelling reasons for taking this report so seriously is that the government of the United States of America took this report so seriously. It is hard to believe that an agency such as the FBI would take and retain reports of flying saucers which had no special meaning for them. Add to this the fact that this report was over ten years old at the time it was taken and that it concerns a report originating in another country.
>
> The FBI operates within the USA and usually does not concern itself with foreign matters unless they have meaning for the internal security of the United States.

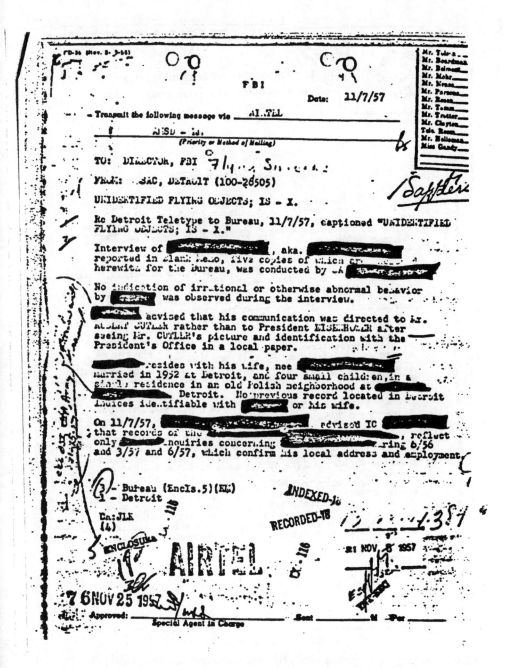

FD-36 (Rev. 3-3-55)

FBI

Date: 11/7/57

Transmit the following message via ___AIRTEL___
_____AISD - id._____
(Priority or Method of Mailing)

TO: DIRECTOR, FBI

FROM: SAC, DETROIT (100-26505)

UNIDENTIFIED FLYING OBJECTS; IS - X.

Re Detroit Teletype to Bureau, 11/7/57, captioned "UNIDENTIFIED
FLYING OBJECTS; IS - X."

Interview of ▮▮▮▮▮▮▮▮▮, aka. ▮▮▮▮▮▮▮▮
reported in blank Memo, five copies of which are ▮▮▮▮▮
herewith for the Bureau, was conducted by SA ▮▮▮▮▮▮▮▮

No indication of irrational or otherwise abnormal behavior
by ▮▮▮▮ was observed during the interview.

▮▮▮▮▮ advised that his communication was directed to Mr.
ROBERT CUTLER rather than to President EISENHOWER after
seeing Mr. CUTLER's picture and identification with the
President's Office in a local paper.

▮▮▮▮▮▮ resides with his wife, nee ▮▮▮▮▮▮▮▮▮▮
married in 1952 at Detroit, and four small children, in a
single residence in an old Polish neighborhood at ▮▮▮▮▮
▮▮▮▮▮▮▮ Detroit. No previous record located in Detroit
Indices identifiable with ▮▮▮▮ or his wife.

On 11/7/57, ▮▮▮▮▮▮▮▮▮▮▮▮▮▮▮ advised IC ▮▮▮▮▮
that records of the ▮▮▮▮▮▮▮▮▮▮▮▮▮▮, reflect
only ▮▮▮▮ inquiries concerning ▮▮▮▮▮▮▮▮ during 6/56
and 3/57 and 6/57, which confirm his local address and employment.

③- Bureau (EncIs.5)(RM)
1 - Detroit

Dn:JLK
(4)

INDEXED-16
RECORDED-16

The first page of a series of 1957 FBI documents concerning a test flight of a Haunebu.

UNITED STATES DEPARTMENT OF JUSTICE

FEDERAL BUREAU OF INVESTIGATION

Detroit 31, Michigan

November 7, 1957

In Reply, Please Refer to File No.

UNIDENTIFIED FLYING OBJECTS

In response to a letter directed by him to Mr. Robert Cutler, Special Assistant to President Dwight D. Eisenhower, reflecting that he "might have some information about the rocket in Texas," ▆▆▆▆ Detroit, was interviewed November 7, 1957, and furnished the following information:

Born February 19, 1926 in the State of Warsaw, Poland, ▆▆ was brought from Poland as a prisoner of War to Gut Alt Golssen, approximately 30 miles east of Berlin, Germany, in May, 1942, where he remained until a few weeks after the end of World War II. He spent the following years at Displaced Persons Camps at Work, Strasburg, Offenburg, Milheim and Freiburg, Germany. He attended a radio technician school at Freiburg and for about a year was employed in a textile mill at Laurachbaden, Germany. He arrived in the United States at New York, May 2, 1951, via the "S.S. General Stewart" as a Displaced Person, destined to the Reverend ▆▆▆▆▆▆, Hamtramck, Michigan; his alien registration number —

Since May, 1951, he has been employed at the Gobel Brewery, Detroit.

News report of mysterious vehicle in Texas causing engines to stall prompted him to communicate with the United States Government concerning a similar phenomenon observed by him in 1944 in the area of Gut Alt Golssen.

According to ▆▆▆▆, during 1944, month not recalled, while enroute to work in a field a short distance north of Gut Alt Golssen, their tractor engine stalled on a road through a swamp area. No machinery or other vehicle was then visible although a noise was heard described as a high-pitched whine similar to that produced by a large electric generator.

"62-83594-284"

The second page of a series of 1957 FBI documents concerning a test flight of a Haunebu.

An "SS" guard appeared and talked briefly with the German driver of the tractor, who waited five to ten minutes, after which the noise stopped and the tractor engine was started normally. Approximately 3 hours later in the same swamp area, but away from the road where the work crew was cutting "hay", he surreptitiously, because of the German in charge of the crew and "SS" guards in the otherwise deserted area, observed a circular enclosure approximately 100 to 150 yards in diameter protected from viewers by a tarpaulin-type wall approximately 50 feet high, from which a vehicle was observed to slowly rise vertically to a height sufficient to clear the wall and then to move slowly horizontally a short distance out of his view, which was obstructed by nearby trees.

This vehicle, observed from approximately 500 feet, was described as circular in shape, 75 to 100 yards in diameter, and about 14 feet high, consisting of dark gray stationary top and bottom sections, five to six feet high. The approximate three foot middle section appeared to be a rapidly moving component producing a continuous blur similar to an aeroplane propeller, but extending the circumference of the vehicle so far as could be observed. The noise emanating from the vehicle was similar but of somewhat lower pitch than the noise previously heard. The engine of the tractor again stalled on this occasion and no effort was made by the German driver to start the engine until the noise stopped, after which the engine started normally.

Uninsulated metal, possibly copper, cables one and one-half inch to two inches in diameter, on and under the surface of the ground, in some places covered by water, were observed on this and previous occasions, apparently running between the enclosure and a small concrete column-like structure between the road and enclosure.

This area was not visited by ███████ again until shortly after the end of World War II, when it was observed the cables had been removed and the previous locations of the concrete structure and the enclosure were covered by water. ███████ stated he has not been in communication since 1945 with any of the work crew of 16 or 18 men, consisting of Russian, French and Polish POWs, who had discussed this incident among themselves many times. However, of these, ███████ was able to recall by name only ███████ no address known, described as then about 50 years of age and presumed by ███████ to have returned to Poland after 1945.

R. OFFICIAL   74

-2-

The third page of a series of 1957 FBI documents concerning a test flight of a Haunebu.

Could the reason that this report was taken and retained for so many years be that it did, in fact, have meaning for the internal security of the United States? Did it have something to do with the flying saucers seen over Texas at the time which also stopped motor vehicles?

As an alternative to the security issues, could there have been another reason that the FBI was so interested in flying saucers? Did the FBI desperately want information on UFOs which was held by the military and other branches of the intelligence community which was not shared with the FBI? It has been rumored that J. Edgar Hoover, head of the FBI at the time, was very interested in learning these secrets but was held "out of the loop." It could be that the FBI was already aware of German saucers through security clearances done on German scientists coming to the USA under Operation Paperclip. The ego of J. Edgar Hoover may have been a factor in the Bureau's quest to learn more on this subject. Hoover may have wanted to be on an equal footing with other intelligence chiefs.

For whatever reason, something in these reports resonated with the FBI. The report was taken seriously, investigated and kept. This fact alone speaks volumes for the existence of UFOs in general and German saucers in particular.[34]

This is a good point Stevens makes about the FBI wanting to know more about the flying saucers that were being reported around the world. He is essentially saying that J. Edgar Hoover felt out of the loop on this saucer stuff and thought the CIA was withholding information from him. Indeed they were. And CIA agents were actively working with postwar Nazis on all sorts of stuff with Operation Paperclip and more. In my book *Antarctica and the Secret Space Program*[10] I presented the evidence surrounding the allegation that the CIA had used a U-boat stationed at a secret submarine base in the Canary Islands for smuggling. The CIA knew a lot more about the postwar activities of the SS, saucers, and secret submarine activities than the FBI knew, and they were not sharing that information with them.

Geheime Kommandosache

Flugkreisel-Erprobung, Stand / Anzahl Erprobungsflüge:

HAUNEBU I   (vorhanden 2 Stück)   52   E-IV
HAUNEBU II   (vorhanden 7 Stück)   106   E-IV
HAUNEBU III (vorhanden 1 Stück)   19   E-IV
(VRIL I)   (vorhanden 17 Stück)   84   (Schumann)

Empfehlung:
Beschleunigen von Abschlußerprobung
und Produktion „Haunebu II"
+ „VRIL I"

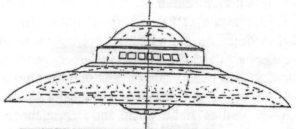

MITTELSCHWERER BEWAFFNETER FLUGKREISEL, TYPE „HAUNEBU I"

Durchmesser: 25 Meter
Antrieb: Thule-Tachyonator 7b
Steuerung: Mag-Feld-Impulser 4
Geschwindigkeit: 4800 Kilom.p.Std. (rechn. bis 17000)
Reichweite in Flugzeit: 18 Stunden
Bewaffnung: 2 x 8cm KSK in Drehtürmen und 4 x MK 108, starr nach vorn
Außenpanzerung: Doppel-Victalen
Besatzung: 8 Mann
Weltallfähigkeit: 60 %
Stillschwebefähigkeit: 8 Minuten
Allgemeine Flugfähigkeit: Tag wie Nacht
Grundsätzliche Einsatztauglichkeit: 60 %
Frontverfügbarkeit: Nicht vor Jahresende .44

Bemerkung: Die RS-E-IV hält Konzentration auf bereits im Versuch
stehende „Haunebu II" für sinnvoller als an beiden Typen parallel
weiterzuarbeiten. „Haunebu II" verspricht entscheidende Verbesserungen
in nahezu allen Punkten. Höhere Herstellungskosten scheinen gerecht-
fertigt – besonders mit Blick auf Führer-Sonderbefehl, Flugkreisel
betreffend.

Plans for the Haunebu I craft.

On top of that, we must remember that military secrets and their sort tend to be kept secret. In most countries, there is only one military and that single establishment controls ground forces, the navy, the air force, and the military intelligence of that country. Only one country has a different structure (at least until recently) than this and this country is the United States of America. The USA has traditionally kept all of the branches of the military—the Army, Navy, Marines, and the Coast Guard—separate from each other. This has meant that the Army, Navy and Marines (the Air Force separated from the Army and retained the same intelligence network) all had their own intelligence networks and operatives, and the CIA and FBI were further, separate intelligence agencies.

Let me repeat, no other country has ever had such a separation of military and intelligence operations than the USA, the most powerful nation on Earth. This is ultimately a good thing, though this intelligence power has been consolidated in the last few decades. Because of this rivalry, John Connelly, the former Governor of Texas, shot during the Kennedy assassination, said in a television interview, "If the Navy had done it the Army would have found out and turned them in." Indeed, cooperation between the Navy, Army, CIA and the FBI was very minimal. Their most closely guarded information and enquiries were kept to themselves. In many cases they deeply distrusted the other agencies. Many books and movies have been made on this premise, too many to name here. These books and movies often feature plots that involve interagency plotting and betrayal and CIA agents—often as drug runners—fighting with other agencies. Sometimes the military intelligence operatives look like the good guys, while the CIA are the bad guys, or the roles can be reversed. The FBI, as portrayed in the television series *The X Files*, is usually in the middle, trying to figure out what is going on in this strange universe of questionable characters.

The Haunebu Secret Files

Henry Stevens says a Haunebu craft is a large-size flying saucer, at least 30 feet or so in diameter, with a classic dome on its upper side, and without any indication of rotating discs, wheels or parts. It is thought to make a whining sound when starting up,

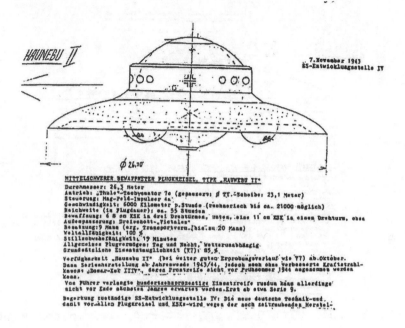

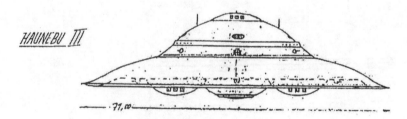

Plans for the Haunebu II and III craft.

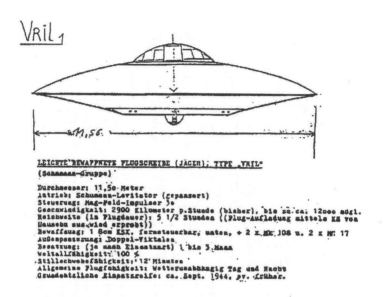

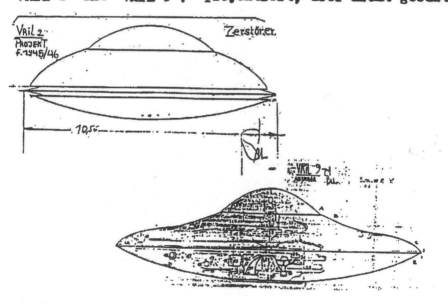

One of the documents containing plans for the Vril craft.

as indicated in the FBI documents described above. Because the craft is electric there is a glow about it and at night the craft may be brightly lit.

It would seem that what is starting up are mercury plasma gyros which are electrified and create a whirling, gyroscopic tornado of energy inside sealed globes or spheres. These spheres are indicated in the Haunebu diagrams as we shall see. These plasma gyros make a whining sound when they first start up and the electromagnetic field that they generate interferes with other electrical devices in the vicinity, including car and aircraft motors. Indeed, this was essentially what the mysterious "foo fighters" did in the last days of the war—they interfered with electrical systems of the bombers and forced them to turn back to England.

The files and photos describe two types of flying saucer, the Vril, a small two-man craft and the Haunebu, a larger saucer-shaped craft that had seats for nine people. A third type of craft is shown, the classic cigar-shaped mothership called the Andromeda. This craft could allegedly hold one Haunebu II craft and four of the smaller Vril craft.

The projects were supposedly under the supervision of the "Vril-Gesellschaft" and of the SS E IV (a secret development center for alternative energy of the SS), and as such they were not directly under Hitler's and the Nazi Party's orders and were not really planned for war use. But later when Germany's situation deteriorated the SS began to think about using the flying disks in the war. They were originally scout craft. Later they were used for other purposes.

Supposedly, the Vril and Haunebu craft had these statistics:

Vril1 from Sept. 44
Diameter 11.5 m
Drive: Schuman levitator (antigravitation eqpm.)
Steuerung/steering: mag-field-impulser
Velocity: 2900-12000km/h
Capacity: 5.5 hrs in air

Haunebu I from Dec. 44
Diameter 25 m

Drive: Thule tachyonator 7b (antigravitation eqpm.)
Steuerung/steering: mag-field-impulser
Velocity: 4800-17000km/h
Capacity: 18 hrs in air
Crew 8 people

Haunebu II from 43-44
Diameter 26.3 m
Drive: Thule tachyonator 7b (antigravitation eqpm.)
Steuerung/steering: mag-field-impulser
Velocity: 6000-21000km/h
Capacity: 55 hrs in air
Crew 9 people

Haunebu III from sometime in 45
Diameter 71 m
Drive: Thule tachyonator 7b and Schuman levitators (antigravitation eqpm.)
Steuerung/steering: mag-field-impulser
Velocity: 7000-40000km/h
Capacity: 8 weeks in air
Crew 32 people

It is interesting to note in the statistics for the Haunebu that the Vril craft could be in the air for five and half hours while the Haunebu I could be in the air for 18 hours. The Haunebu II could be in the air for 55 hours and the Haunebu III could be in the air for an astonishing eight weeks! Clearly, the Haunebu craft needed toilet facilities on them but the Vril craft had no toilet or

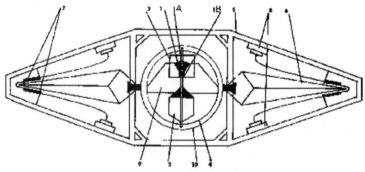

A diagram of the Vril craft.

other facilities. It was apparently noted in some crashed UFOs that there were no toilet facilities of any kind, which indicated to early military researchers that these were not long-haul craft meant for long missions. Still, the amazing speed of the Vril craft would allow it to go long distances even in the relatively short time it could fly.

As in all clandestine programs, the names of the devices or craft were ultimately secret, and the odd name Haunebu would have originally been a code word.

So, what is the meaning of the code word Haunebu? I found it very difficult to find out what this word meant. Unlike the term vril, no explanation of the word Haunebu is given in any of the literature that can be found on the subject. As with many code names German code names did not necessarily have anything to do with the actual mission, device, or operation itself. This may be the case with the word Haunebu—it may have some arcane meaning that only a few of the inner circle of the Vril Society would know.

In looking it up on various dictionaries on the Internet, one lead seemed to show that it was a type of hair color. However, the best definition that I was able to find was this from wiktionary.org:

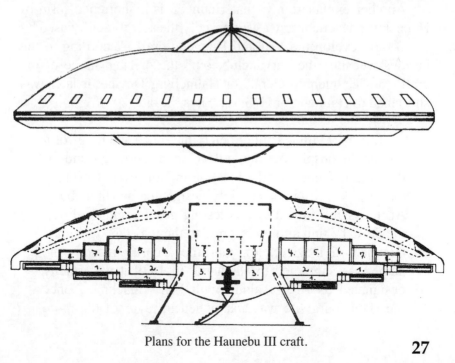

Plans for the Haunebu III craft.

Haunebu (plural Haunebus or Haunebu)

(*Egyptology*, plural "Haunebu") A member of a people from the Aegean Sea.

(*ufology*, plural "Haunebus") Any of a class of flying saucers supposedly built by the Nazis.

So, does the meaning behind the code word "Haunebu" refer to the original Greek raiders known to the Egyptians as the Sea Peoples who fought against Egypt circa 1200 BC and were ultimately repelled? These Sea Peoples were depicted as having horned helmets, like Vikings, and they overwhelmed coastal Egypt for a period until they were defeated. These people essentially settled in the Eastern Mediterranean in what is today Lebanon, Turkey and Greece. These people were, apparently, the "Haunebu." It has been suspected that the Sea Peoples were from Germany and Scandinavia and this could fit in with the SS's penchant for using esoteric and raider-type names such as vril, werewolf, death's head, storm trooper and so on. The SS Death's Head rings, such as the one Himmler wore, are now valuable collector's items and were worn by most of the U-boat officers and other commanders.

Another explanation is that Haunebu is a shortened form of Hauneburg. Hauneburg was supposedly the secret facility used by the SS to develop the Vril and Haunebu craft. According to the curious website "the zurvanclub.com" the original name of the craft was "Hauneburg Gerat" or Hauneburg Device. It later was shortened to Haunebu or H-Gerat. Says the site:

> Since 1935 the Thule Society had been scouting for a remote, inconspicuous, underdeveloped testing ground for their craft. Thule found a location in Northwest Germany that was known as or possibly designated as Hauneburg. At the establishment of this testing ground and facilities, the SS E-IV unit simply referred to them as the "H-Gerat" (Hauneburg Device). For wartime security reasons the name was shortened to Haunebu. In 1939 it was briefly designated as RFZ-5 along with Vril's machines, once the Hauneburg site was abandoned in favor of the more

suitable Vril Arado Brandenburg aircraft testing grounds.

So, this seems to be the origin of the name Haunebu; it is named after the facility it was designed in. The SS often used occult type names and symbols for their ships and aircraft: Vril, Atlantis, Andromeda, Werewolf and the skull and cross bones—a typical Masonic symbol first used by the Knights Templar when their fleet was outlawed by the Vatican. But with the Haunebu, we have a curious and mundane name for what was one of the most top-secret projects during WWII. And, the super-secret nature of the craft meant that it had no name, only the code name "H-Gerat" was used in any communications. Ultimately, plans for the craft had to give the flying saucer a name and this became Haunebu. I must admit, this seems to be a correct origin for the word.

But we still do not know where this "Hauneburg" was located, except that it was supposedly in northwest Germany. There is no town or area called Hauneburg that I am able to find. Apparently this was a secret manufacturing site probably in a rural forested area.

A prehistoric hill fort called the Heuneburg is located by the river Danube in Hundersingen near Herbertingen, in the south of Germany, close to the modern borders with Switzerland and Austria. It is considered to be one of the most important early Celtic centers in Central Europe. It is a fortified citadel and there are extensive remains of settlements and burial areas spanning several centuries. Though it has a similar name, it does not seem that Heuneburg is the location for the facility creating the Haunebu, however, maybe this facility, where ever it was, took its name from this ancient site.

An Armed, Flying Gyro

Supposedly, the Haunebu was an "armed flying gyro" that was first tested in 1939. Some websites claim that working models of the Haunebu were already being built by 1942. According to the Vril websites on the Internet that promote the mysterious Maria Orsic and her contact with the planet Aldebaran, we have this curious timeline that includes mentions of Operation Paperclip and of the secret Antarctica base in Neuschwabenland, sometimes called Point 211:

In August 1939 the first RFZ 5 took off. It was an armed flying gyro with the odd name "HAUNEBU I." It was twenty-five meters across and carried a crew of eight. At first it reached a speed of 4,800 km/h, later up to 17,000 km/h. It was equipped with two 6 cm KSK ("Kraftstrahlkanonen," power ray guns in revolving towers) and four machine guns. It had a 60% space capability. By the end of 1942 the HAUNEBU II was ready. The diameters varied from twenty-six to thirty-two meters and their height from nine to eleven meters. They carried between nine and twenty people, had a Thule Tachyonator drive and near the ground reached a speed of 6,000 km/h. It could fly in space and had a range of fifty-five flying hours. At this time there existed already plans for a large-capacity craft, the VRIL 7 with a diameter of 120m. A short while later the HAUNEBU III, the showpiece of all disks, was ready, with seventy-one meters across. It was filmed flying. It could transport thirty-two men, could remain airborne for eight weeks and reached at least 7,000 km/h (according to documents in the secret SS archives up to 40,000 km/h).

At the beginning of 1943 it was planned to build in the Zeppelin works a cigar-shaped mother ship. The ANDROMEDA DEVICE of a length of 139m should transport several saucer-shaped craft in its body for flights of long duration (interstellar flights).

By Christmas 1943 an important meeting of the VRIL-GESELLSCHAFT took place at the seaside resort of Kolberg. The two mediums Maria Orsic and Sigrun attended. The main item on the agenda was the ALDEBARAN PROJECT. The mediums had received precise information about the habitable planets around the sun Aldebaran and one began to plan a trip there. At a January 22, 1944 meeting between HITLER, HIMMLER, Kunkel (of the Vril Society) and Dr. Schumann this project was discussed. It was planned to send the VRIL 7 large-capacity craft through a dimension channel independent of the speed of light to Aldebaran. According to Ratthofer a first test flight in the dimension channel took place in the winter of 1944. It barely missed disaster, for photographs show the Vril 7 after the flight looking "as if it had been flying for a hundred years." The outer skin was looking aged and was damaged in several places.

On February 14, 1944, the supersonic helicopter—constructed by Schriever and Habermohl under the V 7 project—that was equipped with twelve turbo-units BMW 028 was flown by the test pilot Joachim Roehlike at Peenemunde. The vertical rate of ascent was 800 meters per minute, it reached a height of 24,200 meters and in horizontal flight a speed of 2,200 km/h. It could also be driven with unconventional energy. But the helicopter never saw action since Peenemunde was bombed in 1944 and the subsequent move to Prague didn't work out either, because the Americans and the Russians occupied Prague before the flying machines were ready again.

In the secret archives of the SS the British and the Americans discovered during the occupation of Germany at the beginning of 1945—photographs of the Haunebu II and the Vril I crafts as well as of the Andromeda device. Due to President Truman's decision in March 1946 the war fleet command of the U.S. gave permission to collect material of the German high technology experiments.

Under the operation PAPERCLIP German scientists who had worked in secret were brought to the U.S. privately, among them VIKTOR SCHAUBERGER and WERNHER VON BRAUN.

A short summary of the developments that were meant to be produced in series: The first project was led by Prof. Dr. mg. W. 0. Schumann of the Technical University Munich. Under his guidance seventeen disk-shaped flying machines with a diameter of 11.5 m were built, the so-called VRIL-1-Jager (Vril-1 fighters), that made 84 test flights. At least one VRIL-7 and one VRIL-7 large capacity craft apparently started from Brandenburg—after the whole test area had been blown up—towards Aldebaran with some of the Vril scientists and Vril lodge members.

The second project was run by the SS-W development group. Until the beginning of 1945 they had three different sizes of bell-shaped space gyros built: The Haunebu I, 25m diameter, two machines built that made 52 test flights (speed ca. 4,800 km/h). The Haunebu II, 32m diameter, seven machines built that made 106 test flights (speed ca. 6,000 km/h). The Haunebu II was already planned for series production. Tenders were asked from the Dornier and Junkers aircraft manufacturers, and at the end of March 1945 the decision was made in favor of Dornier. The official name for the heavy craft was to be DO-STRA (DOrnier STRAtospehric craft). The Haunebu III, 71m diameter, only one machine built that made at least 19 test flights (speed ca. 7,000 km/h).

The ANDROMEDA DEVICE existed on the drawing board, it was 139m long and had hangars for one Haunebu II, two Vril I's and two Vril II's.

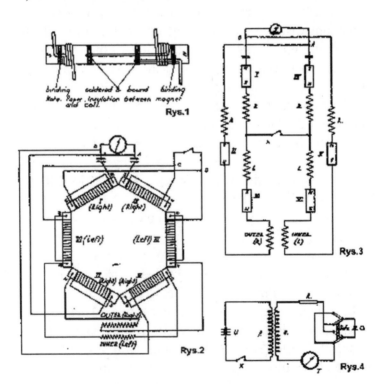

A diagram of the Coler Converter.

There are documents showing that the VRIL 7 large capacity craft was used for secret, still earth-bound, missions after it was finished and test flown by the end of 1944:

1. A landing at the Mondsee in the Salzkammergut in Austria, with dives to test the pressure resistance of the hull.

2. Probably in March and April 1945 the VRIL 7 was stationed in the "Alpenfestung" {Alpine Fortress} for security and strategic reasons, from whence it flew to

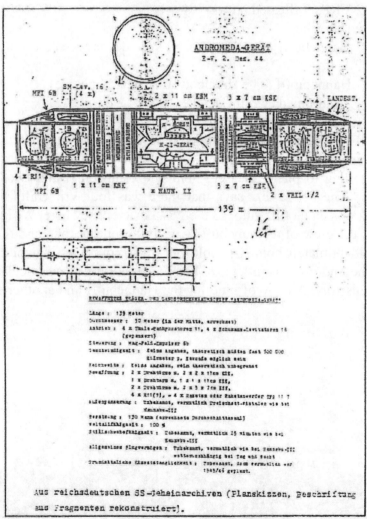

Plans for the Andromeda Craft, a tubular airship that held one Haunebu disk craft and several Vril disk craft.

Spain to get important personalities who had fled there safely to South America and "NEUSCHWABENLAND" to the secret German bases erected there during the war.

From this fascinating list of craft and their statistics, and information coming from the 1989 document dump in Britain that included all of the diagrams that we are seeing, we can gather that at the end of the war there were:

17 VRIL-1-Jager (Vril-1 fighters, seated crew of one)
2 VRIL-7 (Seated crew of ?)
2 Haunebu I
7 Haunebu II (The Haunebu II seated 9 crew, could hold up to 20)
1 Haunebu III (The Haunebu III seated 32 people)
2? Andromeda Craft (Device)

The Andromeda craft, called a device in the Vril material on the Internet, was said to be a 139-meter-long tubular craft that, according to plans released, held one Haunebu II and two Vril craft. The above text says that there was four Vril craft in the Andromeda mother ship, so there is a discrepancy here. We know that the crew of a Haunebu II was up to nine people, and that it could ultimately hold 20 people, but we do not know what the crew of the Andromeda craft would be. We might surmise by calculating in this way: A crew of nine for the Haunebu II and another four

A photograph of a Haunebu II craft on the ground.

A photo of a Vril craft in flight.

crewmembers for the smaller Vril craft. If there was another 20 possible crewmembers we get a figure of 33 crewmembers.

This is an interesting number for a secret society of Teutonic Knights who believed that they were fighting a World War against an English-French Masonic society who believed that they were descended from the Knights Templar. The Templars, after their suppression, founded the first modern banks and courts at the Old Bailey in London's Old City. The underground stop here is Temple, and the original circular Templar church in London can be found here as well. The Teutonic Knights were a third branch of the Crusaders, with the Knights of St. John and the Knights Templar being the other two branches. The Nazi SS fancied themselves as the reincarnations of the Teutonic Knights of old—a group that was allied with the British, French and Templars, but ultimately opposed to the Russians and other Slavic forces.

In many ways World War II was about secret societies fighting for control of the world in the 1930s and 1940s. There was still time for the Germans and Japanese and their allies to grab more control of the world, and the Thule Society, the Vril Society and the SS all preached the message of an expansive Germany—a Germany that was all over the world, like the British Empire. But their nemesis was the British Empire, a Masonic-Templar establishment that had its own secret societies. It is important to remember that secret societies like the Masons, Knights of Columbus, and many others were all the rage until television came along in the 1940s and 1950s. The British were actually seen by Hess and other members

A photo of a Haunebu II craft on the ground.

of the Thule Society as allies, not enemies. However Churchill and the British military would not negotiate on any terms with Hitler's regime.

We will see in the next two chapters that Hitler, the Nazis and the SS did not consider Britain, France and the West as their enemies—merely as their adversaries. The real goal of the Germans was to push eastward into Russia and the Ukraine and to capture the oilfields at Baku in Azerbaijan. We will discuss this in the next chapter. Germany did not have any oilfields of its own and was constantly looking for sources of energy. This is why the electrical craft predicted by Nikola Tesla were of such interest to the Germans—they needed aircraft and other equipment that did not use petroleum products as their main source of power. Thus the intensified research in "electric spacecraft." Tesla had provided the platform and the Germans used their money and technology to build the first of what could be described as "Tesla Craft."

Armed Flying Saucers and the Viktalen Metal

While the documents from the Third Reich released in 1989 do not state that the Andromeda craft was ever built, photos of the craft apparently exist. This seems to be the craft of the Adamski photos and many others. It seems difficult to believe that there was only one of these craft in existence, but that may be the case.

It seems quite possible, however, that a number of these craft were produced at the end of the war—and in the years after—in the various secret manufacturing facilities in Germany, Antarctica, South America, Greenland and elsewhere. The Germans were famous for shipping parts for various airplanes to all corners of the world so they could be assembled when the time was right. Some of the stories even say that the Germans had the parts for long-range aircraft at the base(s) in Antarctica but they were never able to fly for the lack of certain parts.

The curious website "the zurvanclub.com" claims that the Haunebu was first flown in 1939. It also claims that a special alloy metal called Viktalen was used in the craft. Says the site:

> The weapons developed by the black operations science divisions of the Nazi party led to many 'radical' aircraft experimentations, including the 'One Wing' Horton planes of 142-foot wingspan submitted for approval in 1944, which would have been able to fly from Berlin to NYC and back without refueling, thanks to the blended 'one wing' design and six BMW 003A or eight Junker Jumo 004B

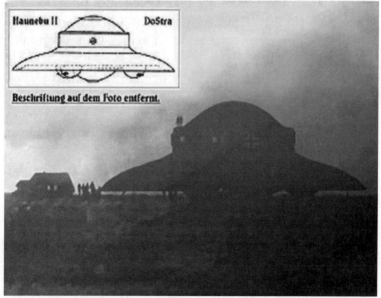

A photo of a Haunebu II craft on the ground.

A photograph of a Haunebu II in flight with a cannon on the underside.

turbojets.

Also, the first rocket airplanes and jet airplanes such as the Messerschmitt Me 163 Komet and Messerschmitt Me 262 were fully developed and functional. However, German UFO theories describe successful attempts to develop advanced aircraft and spacecraft prior to and during World War II, and further assert the postwarsurvival of these craft were carried on in secret to underground bases in Antarctica.

Allegedly, the first model was called JFM "Jenseitsflugmaschine" or "Other World Machine", that was followed by the RFZ "RunFlugZeug" or the "Round Aircraft" series that finally derived in the Vril and Haunebu projects…

Haunebu I. First allegedly flown in 1939. The craft of which two prototypes were constructed were 25 meters in diameter, had a crew of eight and could achieve the incredible initial velocity of 4,800 km/h, but at low altitude. Further enhancement enabled the machine to reach 17,000 km/h.

Haunebu II do-stra. The Dornier STRAtospharen Flugzeug/Stratospheric Aircraft or Haunebu II, was

A modern model for the Haunebu II craft.

tested in 1944. Two prototypes were built. These massive machines, several stories tall, were crewed by 20 men. They were also capable of hypersonic speed beyond 21,000 km/h.

Haunebu III. Yet larger still was the 71-meter diameter lone prototype that was constructed before the close of the war. It was crewed by 32 men and could achieve speeds of 7,000 to 40,000 km/h. It had a triple Viktalen hull and it's said to have had a flight endurance of 7 to 8 weeks. Allegedly, the craft made 19 test flights.

Vril 1. As war had started in 1939 the Haunebu I became the RFZ-5 and by 1941 the RFZ-7 had become the Vril 1 Jager (Hunter). The reason for the changes was due to Thule's revolutionary Triebwerk (Thrustwork) engine that used rotating electro-magnetic-gravitational fields to affect gravity. Vril had by 1941 perfected the SM-Levitator as well and thus two new series entered limited construction, but with slightly different goals. The first purely Vril disc—the Vril 1 Jager (Hunter) was constructed in 1941 and first flew in 1942. It was 11.5 meters in diameter, had a single pilot, and could achieve 2,900 km/h - 12,000 km/h! It flew with a metal dome at first but subsequent test

39

versions had a heavily reinforced glass dome and could seat two crewmembers. Flight endurance was 5.5 hrs. It was planned to arm this craft with two MK-108 cannon plus 2 MG-17 machineguns. Seventeen of these craft were constructed and tested between 1942-44 with 84 test flights. The Vril 2 Zerstorer (Destroyer) was a highly advanced oval shaped disc that was much too complex for the time period; thus it was projected for 1945/46, so no construction was started. The Vril 3 and 4 may have been photographed but no surviving information is found on them while Vril 5 and 6 likewise do not show up and may have only been projects.

Vril 7. The Vril 7 Geist (Ghost) was 45 meters in diameter and crewed by fourteen men. It was built in 1944 and tested at Arado-Brandenburg using Vril's own

A photo of a Vril craft in flight very near a road and car.

engine. Vril's medium Sigrun made frequent trips to the facility to oversee construction and testing. In 1944, Arado engineers approached her with a request. They wanted to know if the Vril Triebwerk could be adapted to one of their projects—the Arado E.555 strategic bomber. They were abruptly told, "No" and returned to their designs which resulted in eleven different versions of the bomber. Sigrun was actually insulted because the entire purpose of the Vril discs was aimed at space flight. No conventional bomber could withstand the heat of the velocity achieved by these machines which were constructed of hulls specially made of an advanced metal called Viktalen (in some sources Victalen or Viktalon). The Vril 1 had a single hull of this type, the Vril 7 two. The large Haunbeu III had three! With the SS supervising all aspects of the disc programs every model had to have at least theoretical provision for armament. In the Vril 7 Geist it would have been four MK-108 cannon.

Vril Odin. The Vril 8 Odin was the last official Vril disc that was flight tested in the spring of 1945 during the collapse. This disc had an automatic Oberon upward-firing gun installation on top of the control center.

The Vril Gesellschaft had started evacuating to Base

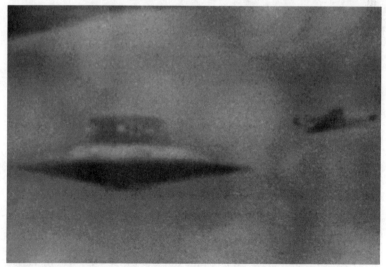

A photo of a Vril craft in flight with an Me-109 in the background.

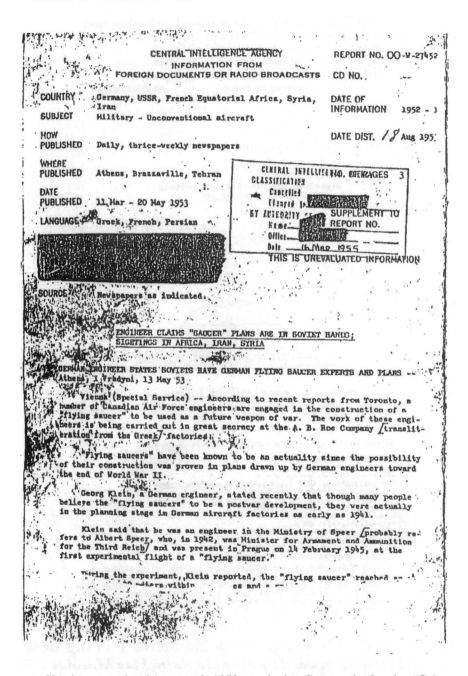

CENTRAL INTELLIGENCE AGENCY   REPORT NO. OO-W-27452
INFORMATION FROM
FOREIGN DOCUMENTS OR RADIO BROADCASTS   CD NO.

COUNTRY Germany, USSR, French Equatorial Africa, Syria,
Iran   DATE OF
SUBJECT   Military - Unconventional aircraft   INFORMATION   1952 -)

HOW   DATE DIST. 18 Aug 195:
PUBLISHED   Daily, thrice-weekly newspapers

WHERE
PUBLISHED   Athens, Brazzaville, Tehran   CENTRAL INTELLIGENCE. COVERAGES 3
                                                CLASSIFICATION
DATE   Cancelled
PUBLISHED   11 Mar - 20 May 1953   Cleared by
                                                BY AUTHORITY   SUPPLEMENT TO
LANGUAGE   Greek, French, Persian   Kine   REPORT NO.
                                                Office
                                                Date   16 Mar 1955
                                                THIS IS UNEVALUATED INFORMATION

SOURCE   Newspapers as indicated.

ENGINEER CLAIMS "SAUCER" PLANS ARE IN SOVIET HANDS;
SIGHTINGS IN AFRICA, IRAN, SYRIA

GERMAN ENGINEER STATES SOVIETS HAVE GERMAN FLYING SAUCER EXPERTS AND PLANS --
Athens, I Vradyni, 13 May '53

Vienna (Special Service) -- According to recent reports from Toronto, a
number of Canadian Air Force engineers are engaged in the construction of a
"flying saucer" to be used as a future weapon of war. The work of these engineers is being carried out in great secrecy at the A. B. Roe Company [translit-
eration from the Greek] factories.

"Flying saucers" have been known to be an actuality since the possibility
of their construction was proven in plans drawn up by German engineers toward
the end of World War II.

Georg Klein, a German engineer, stated recently that though many people
believe the "flying saucers" to be a postwar development, they were actually
in the planning stage in German aircraft factories as early as 1941.

Klein said that he was an engineer in the Ministry of Speer [probably refers to Albert Speer, who, in 1942, was Minister for Armament and Ammunition
for the Third Reich] and was present in Prague on 14 February 1945, at the
first experimental flight of a "flying saucer."

During the experiment, Klein reported, the "flying saucer" reached --
meters within   es and a --

A CIA document dated August 18, 1953 mentioning Germany's planning "flying saucers" as early as 1941. US Navy intelligence probably knew about this, but they had never told the CIA. Even the director of the CIA kept this a secret.

211 in Neu Schwabenland Antarctica in March 1945 so it would seem like the Vril 8 Odin was the last Vril disc actually tested. However, some weeks after Germany surrendered both Haunebu and Vril craft were spotted in the skies over occupied Germany. Although the Vril 9 Abjager (Universal Hunter) was shown as a design on paper, a craft identical to it was photographed post-war.

The site goes on to say that the lone Haunebu III was said to be used extensively in March 1945 to evacuate Vril Society members, many of them in the SS, to secret sites outside of Germany. These

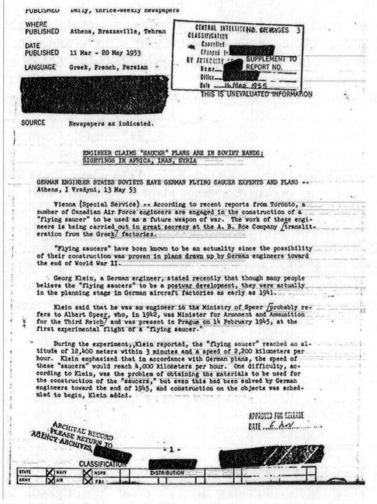

The 1953 CIA document opposite but with the lower portion complete.

were the so-called secret cities of the Black Sun. The metal called Viktalen is interesting to note and the observations in UFO literature of a strange metal are numerous. Indeed, metallic alloys comprise their own special field and such metals as titanium, chromium, gold, platinum and others in special alloy mixtures are just the thing that the Germans would have taken a special interest in.

Submarines That Can Fly

It is impossible to know just what occurred during the war with flying saucers and the technology involved. Many researchers believe that the Vril and Haunebu craft are real but that much of the material on the Internet is some form of disinformation, possibly aimed at neo-Nazi groups. Still, clearly something was going on as far as German saucer research during the war and it must have had something to do with the explosion of UFO activity immediately after WWII. While extraterrestrials may be visiting Earth with advanced craft of their own, at least some of the flying saucers seen and photographed since 1945 must be German craft such as the Haunebu, Vril and Andromeda ships. Some of these craft, particularly those seen in the USA may have been made by the Americans, perhaps beginning in the late 1950s or early 1960s.

It would seem that when Germany officially surrendered in May of 1945 whatever craft were able to fly did so and flew out of Germany to safe havens in Tibet, Spain, Spanish Sahara, Antarctica and South America. Meanwhile, elements of the Third Reich—overtly and covertly—moved their operations to South America and, in a sense, Antarctica. Operations in Germany and other parts of Europe did not come to a complete halt either.

At the end of the war much of the Black Fleet (up to 100 submarines still operating after the war), long-range aircraft, and now flying saucers, were under the control of the SS. These craft often had the Black Sun symbol on them, rather than a swastika or a German Cross. Many of these SS members were formerly of the Thule Society or the Vril Society and they did not surrender at the end of the war.

Like any secret society that comes under attack and is threatened with elimination, they did what they originally did—

they become a secret organization again and operated accordingly. They were not destroyed or dissolved—they still functioned. They still had small bits of territory—secret submarine bases in Greenland, the Canary Islands, Antarctica and South America. They also had secret airfields and airbases in Tibet, Argentina, the Spanish Sahara and elsewhere. They owned an island in the middle of a lake in Patagonia and entire cities in Chile, Paraguay and Argentina were made up of German-speaking immigrants, some of them former SS officers from the Eastern Front.

Henry Stevens says that after the war ended the Nazi International-Third Power set up scientific workshops in South America for the research and development of high technology. This included experimentation and production of flying discs. This coincided with the golden age of flying saucers in the late 1940s through the 1960s. Flying saucer accounts in and around South America abounded, and even today the attitude toward UFOs in all of the countries in South America is one of casual acceptance. The local newspapers and television stations have no problem reporting on UFO activity, and it has been a common topic of discussion since the end of WWII.

Stevens notes how a strange FBI report claimed a German inventor, living in South America, had discovered a new principle of "aerodynamics" and was trying to market this idea. His concept was not one based upon free energy in that he stated his method involved using motorcycle or auto engines. These engines produce the movement of a spinning shaft, whose speed and torque can be harnessed for many uses but at its core is rotary motion.

Stevens says that the modern era of inquiry into the high technology of Nazi Germany only began in the middle to late 1980s and then only hesitantly. One of the first to delve in was the German writer O. Bergmann who published two short books[51] propounding the idea that later German flying discs were field propulsion powered and that there is absolutely no difference between a flying saucer and a flying U-boat at this point. He says they are exactly one and the same, and that these devices made their first appearance at the very end of the Second World War.

Stevens says that Bergmann bolstered his argument with the now familiar "Vril" and "Haunebu" diagrams, in fact he was the

45

first to publish them, and with testimony from all over the world, covering decades of mystery submersibles and UFOs entering and leaving the water. A passage from one of Bergmann's books bears repeating:

> At sometime and someplace, at secret U-boat bases outside the German motherland, U-boats of the German Navy must have been weighed out, and, during the great withdrawal in April/May 1945, missing U-boats must have been equipped with new revolutionary technology and also must have been converted with electromagnetic propulsion. With this, those [U-boats] may [have been] arranged with the same possibilities and technologies as the German flying discs [called UFOs].
>
> UFOs or USOs are certainly not only observed within the seas and oceans and rivers, on the contrary also occasionally diving and surfacing on inland lakes, yes, even in ponds. We need not bother ourselves any more concerning the confusion on differing reports, if it is to be called only UFO, USO or still U-boat because [John] Keel had it completely correct, with the same electromagnetic propulsion it can be what it wants to be, the initial water craft would be able to operate in the air, as other flying craft, as well as the water. The two mediums, air and water, are interchangeable if one is fitted with this phenomenal propulsion, concerning which we have already more closely read in Hugin-Schrift: 'Geheime Wunderwaffen' Bd. III. They could maneuver mutually well in the air or water.[51]

Anti-Gravity or "Field Propulsion"

This anti-gravity or "field propulsion" is apparently a relatively simple technology that is basically electric in nature, rather than jet or rocket powered. Says Stevens about the technology:

> The first clues come from the two scientists who did a ground floor evaluation of Karl Schappeller, his home, his home area, witnesses and the Schappeller technology

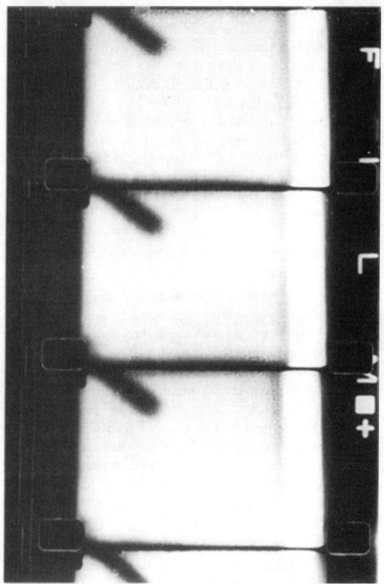

A film strip of a cigar-shaped craft over Victorville, CA in 1953.

itself. Their conclusion was that the technology involved was first introduced by Tesla, and that Schappeller experimented exclusively with Tesla ideas. So does this mean we ought to be looking for clues to Haunebu secrets by looking at old Tesla stuff? Why not? It already seems that Tesla has been involved in the smaller type of field propulsion German flying disc.

...But from Schappeller this technology clearly found its way to the researchers at the Reichsarbeitsgemeinschaft (RAG), a governmental organization charged with energy independence for Germany at a time between the two World Wars. So the RAG was primarily not interested in levitation but nevertheless freely commented on this as being contained among the properties of the device they were building.

Steven then quotes from the Reichsarbeitsgemeinschaft:

The new dynamic technology will enable, in the future, electric locomotives and automobiles to be made without expensive armatures and through circuit-connection to the atmospheric voltage-net to be pushed forward. Prerequisite is that, all things considered, the installation of many sufficient amplification facilities (centers) which transmit the specific "magnetic impulse" given from the Ur-machine to the dynamic globe-elements. Novel aircraft with magneto-static propulsion and steering, which are crash proof and safe from colliding with each other could be built for a fraction of the cost of today's aircraft—and without lengthy training of every person attending (the machines).

Stevens continues by discussing the curious globes that are often seen on the bottom, and sometimes the sides, of the craft:

In conjunction with their discussion on energy production and levitation of aircraft the RAG describes how this all looks to the observer. They propose using seven globes, five surrounding one in the center with one positioned above this center globe. The five feed the center globe and when it reaches some point of saturation, specific magnetic impulses are sent to the seventh. This is what they describe as an Ur-machine. The reader may recall from depictions of the Haunebu machines that there are sometimes three globes visible and presumably one or two larger ones within the body of the craft. The three smaller globes may have an additional function in steering

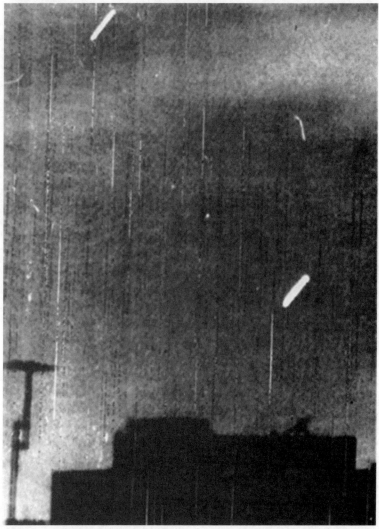

A photograph of cigar-shaped craft over Buenos Aires, Argentina in 1965.

the craft but they also may interact with the globes in the body of the craft to produce magnetic reconnection and the lift producing positrons that result.

Along with the Haunebu type of saucer, this technology would also be suitable for the cigar-shaped flying craft as well as the similarly shaped underwater, unidentified submerged objects. Just as would the two counter-rotating magnetic wheels of the previous homopolar generator inspired devices. For a submarine, the engine should probably be mounted in the front of the ship or perhaps one at the front and one at the rear for forwards and backwards motion. In other words William R. Lyne, all those old German sources, D. H. Haarmann and O. Bergmann are all probably correct, U-boats can fly, sometimes.

A painting of a Haunebu II craft on the ground from one of the models.

This last statement is profound, and as already stated, this electric field technology is essentially able to work underwater as it does in the air. Therefore, incredibly, with this technology the Germans could literally turn submarines into airships, hence the elongated, cigar-shaped UFOs, dark in color, that were commonly seen and even photographed starting immediately after WWII.

Where would the switchover from the old technology to the new one take place in order to turn these U-boats in airships? Probably at the submarine base in Antarctica, and possibly at the bases in the Arctic. Supplies could be brought to the Antarctic base from Argentina and even from southern Africa. Most of the original food and other supplies would have come from Europe.

Though Germany was defeated, the secret SS laboratories, with their black U-boats, long-range aircraft, and even flying saucers, continued to repair older craft but also to build new craft, mainly the flying disks, like the Haunebu. South America became a hotbed of UFO reports for decades and it would appear that many of these craft were assembled in the southern parts of South America, namely Argentina and Chile. Paraguay is another country that had a pro-German stance right up through the 1980s and today. A large ranch in western Paraguay could have all the privacy and support that might be needed to run a secret airbase. Curiously, the Bush family has a large ranch in western Paraguay. Large ranches with airstrips can be makeshift airbases for a short time, and both Lyndon Johnson and George W. Bush had long airstrips on their ranches, where even cargo planes and Air Force One landed.

The Secret Base in Tibet

In my previous book, *Antarctica and the Secret Space Program*, I had an entire chapter on the secret Haunebu base in Tibet. This little-known Nazi base was apparently based in western Tibet during the late 1930s and 40s.

British writer Christopher Hale, in his book *Himmler's Crusade: The Nazi Expedition to Find the Origins of the Aryan Race*,[37] says that Himmler was fascinated by Asian mysticism and therefore wished to send an expedition to Tibet under the auspices of the SS Ahnenerbe. The leader of the expedition was Ernst

Schäfer, a German explorer, hunter and zoologist in the 1930s, specializing in ornithology. He was a scientific member in the Ahnenerbe and held the rank of an SS-Sturmbannführer. Himmler desired that Schäfer perform research based on Hanns Hörbiger's pseudoscientific theory of "Glacial Cosmogony" promoted by the Ahnenerbe. However, Schäfer had scientific objectives and therefore refused to include Edmund Kiss, an adept of Hörbiger's theory, in his team and required 12 conditions to ensure scientific freedom.

Himmler was agreeable to the expedition going ahead provided all members joined the SS including Schäfer. The expedition is widely known as the SS Expedition to Tibet 1938-39.

Christopher Hale observes that "while the idea of 'Nazi botany' or 'Nazi ornithology' is probably absurd, other sciences are not so innocent—and Schäfer's small expedition represented a cross-section of German science in the 1930s." To Hale, this has considerable significance as "under the Third Reich anthropology and medicine were cold-bloodedly exploited to support and enact a murderous creed."[37]

There are strong allegations, though never officially acknowl-

Expedition members with hosts in Gangtok, Sikkim are (from left to right) unknown, unknown Tibetan, Bruno Beger, Ernst Schäfer, Sir Basil Gould, Krause, unknown Tibetan, Karl Wienert, Edmund Geer, unknown, unknown. 1938.

edged, that one of the expedition's purposes was to determine whether Tibet was the cradle of the Aryan race. This was largely done by the taking of cranial measurements and the making of facial casts of local people by the anthropologist Bruno Beger.

Ernst Schäfer, leader of the expedition.

Hale relates the existence of a secret warning issued by propaganda minister Joseph Goebbels to German newspapers in 1940 saying that "the chief task of the Tibet expedition," was "of a political and military nature" and "had not so much to do with the solution of scientific questions," adding that details could not be revealed.[37]

Obviously, the expedition was caught up in the politics of its time, and whenever the SS was involved there was likely some military and technological purpose to an expedition as well. It was suspected at the time that the Germans were scouting out a possible location for an airbase located somewhere on the plateau of Tibet. They could use this airbase to attack the British army in India, which was otherwise beyond the scope of the German air force.

According to Wikipedia:

> Chinese journalist Ren Yanshi, quoting the Austrian weekly *Wochenpresse*, writes that the first major task of the expedition was "to investigate the possibility of establishing the region as a base for attacking the British troops stationed in India" while its second major assignment was "to verify Heinrich Himmler's Nazi racial theory that a group of pure-blooded Aryans had settled in Tibet."

According to American journalist Karl E. Meyer, one of the expedition's aims was to prepare maps and survey passes "for possible use of Tibet as a staging ground for guerrilla assaults on British India."

After traveling by boat from Europe to Calcutta, the SS expedition team assembled in Sikkim's capital of Gangtok, at that time a British protected mini-state like Bhutan. The expedition assembled a 50-mule caravan and searched for porters and Tibetan interpreters. Hale reports that at Gangtok, the British official, Sir Basil Gould, observed them, describing Schäfer as "interesting, forceful, volatile, scholarly, vain to the point of childishness, disregardful of social convention," and noted that he was determined to enter Tibet regardless of whether he got permission.[37]

The team began their Tibetan journey on June 21, 1938, traveling as a caravan through the Teesta River valley and then heading north. In August 1938, a high official of the Rajah Tering, a member of the Sikkimese royal family living in Tibet, entered the team's camp. Schäfer met with the official, and presented him

Under SS pennants and a swastika, the expedition members are entertaining some Tibetan dignitaries and the Chinese representative in Lhasa; left: Beger, Chang Wei-pei Geer; in the center: Tsarong Dzasa, Schäfer; right: Wienert, Möndro (Möndo).

with mule-loads of gifts.

In December 1938 the Tibetan council of ministers in Lhasa formally invited Schäfer and his team to Tibet, but they were forbidden to kill any animals during their stay, due to religious concerns. After a supply trip back to Gangtok, Schäfer learned he had been promoted to SS-Hauptsturmführer, and the rest of the team had been promoted to SS-Obersturmführer because Himmler was so excited about the expedition having successfully entered Tibet.

The team reached the capital of Lhasa on January 19, 1939. Schäfer proceeded to pay his respects and offer gifts to the Tibetan ministers and a nobleman. He also gave out Nazi swastika pennants, explaining the reverence shown for the shared symbol in Germany. They demonstrated their high-quality German rifles and pistols to the Tibetans. The expedition's permission to remain in Lhasa was extended, and Schäfer was permitted to photograph and film the region. The team spent two months in Lhasa and the immediate area, collecting information on the culture, agriculture, and religion. One might also think that they were looking for a suitable place to build an airfield, and a hangar that was inside a cliff or mountain.

Schäfer met the Regent of Tibet, Reting Rinpoche, on several occasions. During one of their meetings, the Regent asked him point blank whether his country would be willing to sell weapons to Tibet.

In March 1939, the expedition left Lhasa, heading for Gyantse escorted by a Tibetan official. After exploring the ruins of the ancient deserted capital city of Jalung Phodrang, they reached Shigatse, the city of the Panchen lamas, in April. They received a warm welcome from the locals, with thousands coming out to greet them. Here, Schäfer claims to have met "the pro-German regent of Shigatse" (the 9th Panchen Lama had died in 1937 and the 10th was not to arrive before 1951). All in all the Germans were made to feel welcome in Tibet and the Tibetans seemed to like the idea of being allies with the Germans, who might aid them in their efforts to keep the British and Chinese from taking too much control of the country, especially in terms of foreign relations with outside countries. As mentioned previously, the Reting Rinpoche

The Thirteenth Dalai Lama in 1932.

had asked if they could buy weapons from the Germans. In fact, these weapons would probably just be given to the Tibetans, along with a German SS unit to administer the airbase and arms depot.

In May of 1938, the expedition returned to Gyantse and Lhasa and began negotiations with local British officials about the trip back to India and transport of the expedition's gear and collections. After Schäfer read a letter from his father who reported to him about the imminent threat of war, and urged him to return to Germany as quickly as possible, Schäfer decided to depart immediately from Lhasa.[37]

After being given two complimentary letters, one to Hitler and the other to Himmler, Schäfer and his companions left. They also took with them two presents for Hitler, those being a Lama dress and a Tibetan hunting dog, as well as a copy of the Tibetan "Bible," the 120-volume *Kang Shur*. The expedition headed south over the Himalayas to Darjeeling and arrived in Calcutta in July of 1939. After a brief stay the Germans boarded a British Airways seaplane at the mouth of the Hooghly River, and began the journey home via Karachi and Baghdad.

Schäfer kept meticulous notes on the religious and cultural customs of the Tibetans, from their various colorful Buddhist festivals to Tibetan attitudes towards marriage, rape, menstruation, childbirth, homosexuality and masturbation. In his account of Tibetan homosexuality he describes the various positions taken by older lamas with younger boys and then goes on to explain how homosexuality played an important role in the higher politics of Tibet. There are pages of careful observation of Himalayan people engaged in a variety of intimate acts. Schäfer presented the results of the expedition on July 25, 1939 at the Himalaya Club Calcutta.

Throughout his stay in Lhasa, Ernst Schäfer had remained in touch with Germany through mail and the Chinese Legation's radio. Heinrich Himmler was reported to have followed the expedition very enthusiastically, writing several letters to Schäfer and even broadcasting a Christmas greeting to him via shortwave. One must presume that, no matter how scientifically pure Schäfer's motives were, the SS would have put several special officers, perhaps unknown to Schäfer, into the team to look for possible airfields and hangars.

Tibetans in Berlin

According to a number of sources, on April 25, 1945, a group of Russian soldiers entering Berlin discovered the dead bodies of six Tibetan monks in a circle, with the body of a seventh monk in the center. The dead man in the center was wearing a pair of bright green gloves. One of the Russian soldiers was a Mongolian and he literally freaked out, dropping to his knees and praying to, or perhaps for, the dead monks. Over the next few weeks the bodies of hundreds of Tibetan monks were found in the ruins of Berlin; all were said to have committed ritual suicide.

What were these monks doing in Berlin and how did they get there? It would have been difficult for these Tibetans to have come to Germany by sea as they would have had to travel through British India and take a ship from a British-held port. Nor could these Tibetans have come overland through China or Russia (via Mongolia) as these countries were at war (with Germany, in Russia's case). So it has been theorized that these Tibetan lamas were flown from Tibet to Germany with all of their religious artifacts and costumes.

This would mean that when the SS returned to Germany they had made plans to make air contact with Tibet and build an airbase on the Tibetan plateau somewhere. I would suggest in western Tibet, possibly even near Mount Kailash, or further west, where the land can be flat and dry with snowy peaks surrounding these

A map showing the possible locations of the secret SS airbase in Tibet marked with the Black Sun.

large, flat valleys. There are many places in western Tibet where only minimal work would need to be done to create a long flat landing spot where a large airplane could safely land. The distance from this area to Austria is approximately the same as from France to New York City, and these flights were meant to be non-fueled return flights. Therefore, a trip to Tibet from Austria would have been a round-trip event with no need for refueling at the new airbase in Tibet.

And the flight could have contained barrels of fuel to be stockpiled for the future need to refuel an airplane while at the airbase in Tibet. Future flights could have continued to add to this cache until a considerable stockpile of fuel (and spare parts) existed in this remote part of Tibet. It seems probable that a stockpile was accumulated that could have refueled a number of these thirsty aircraft. However, such a stockpile would not last forever, nor would an isolated base in Tibet.

Dr. Joseph P. Farrell gives us a rather rational description of what might be considered a German base in Tibet in his 2008 book *Nazi International*.[12] Says Farrell:

The likely possibility was sometime around the period of the massive German offensive on the southern Russian Front in the summer of 1942. It is a little known fact that the Wehrmacht sent armed long-range reconnaissance units as far east as Astrakhan on the northern Caspian Sea, but beyond sowing confusion behind Russian lines, there was little military value in such an operation. [Here Farrell cites the German military historian Paul Carrell in the book *Hitler Moves East: 1941-1943*] However, it is possible that commando-engineered units of the Waffen SS might have been included in these units, and, once the units had penetrated as far east as was operationally safe, the commando units would have been released to continue onward through Kazakhstan and on into Tibet, utilizing Nazi intelligence contacts in these Muslim areas to smooth their passage through hostile territory. Once in Tibet, a makeshift airfield could have been constructed able to handle long-range aircraft. The Tibetans would have then been flown back to Germany. While a hazardous enterprise to be sure, it would have been far safer than attempting to smuggle monks out through British India and back to Germany via U-boat, which, in any case, would have required more trips than would have been required by even one of the large Junkers 290s or the enormous Junkers 390s. As for the purpose of such an operation, this is relatively transparent, for such expertise would have been required if Himmler's SS was to translate its copy of the *Kang Shur* completely and accurately.

This requires some commentary. Since the 1938-1939 Schaefer Expedition is known not to have brought out large numbers of Tibetan monks, but only a copy of the *Kang Shur*, the question becomes, when were these monks smuggled out of Tibet into Nazi Germany, by what route, and for what purpose?[12]

Yes, this is all very possible. Farrell seems to think that the SS might have been able to continue overland from Kazakhstan to Tibet to establish an airfield, but it seems more likely that any travel into Tibet was by aircraft, and Kazakhstan may have been a good location to launch the initial flights into Tibet. Weapons plus other supplies for a base were flown into Tibet, I suggest western Tibet, and a makeshift airfield was created for long-range aircraft.

What I propose happened is this: after a weapons deal with Tibetan ministers in Lhasa was concluded, several SS members of the SS Expedition stayed behind, or had secretly met other SS officers in Lhasa who had come via China.

These SS officers that were left behind in Lhasa were then given a small Tibetan unit, and along with a transmitter that they brought with them from Germany, went to western Tibet to find a suitable airbase for a long-range German aircraft to land. This small team of perhaps two Germans and ten Tibetans could have set up an airfield and beacon that would allow these long-range aircraft to fly from Austria, or more likely closer airfields such as those in Romania (an ally of the Third Reich and not an occupied country) all the way to Tibet, carrying fuel, weapons, generators,

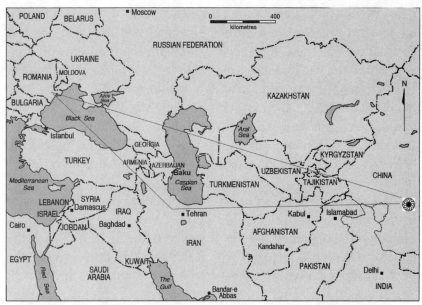

A map showing two routes to Tibet from Romania, one direct and one flying south
of the Caspian Sea. The Black Sun symbol marks the Tibetan airbase.

lights, and more SS officers. Prior to 1944, when the very long-
range aircraft first became available, they would have had to fly
out of airbases in the captured sections of the Ukraine starting in
June of 1941 after Operation Barbarossa. It seems that flights from
Romania or the Ukraine could have made the journey non-stop to
Western Tibet, carrying extra drums of fuel.

After several such early flights bringing machinery and
supplies, a fairly large number of Tibetan lamas could have been
brought to Germany for special use by the SS. Translating the
Kang Shur texts into German would have been one of their duties.
Supplying the Tibetans with brand new German rifles of even the
most basic type would fulfill Germany's "promise" to arm Tibet.
All of this was happening about 15 years before the Dalai Lama,
as a young boy, was to flee Tibet as the Chinese took over the
country.

A curious tale is told by the American doctor Howard Buechner
in his 1991 book *Emerald Cup—Ark of Gold*,[30] of a flight to
Tibet from Austria, late in the war. But first some background
information.

In the summer of 1931 a German mystic and SS officer

61

named Otto Rahn went to the Languedoc area of southern France on a secret mission for Heinrich Himmler and the Nazi SS. His destination was the Cathar fortress of Montségur and his mission was to find the lost treasure from Solomon's Temple, including the Holy Grail as an emerald-encased gold cup. Otto Rahn was said to have discovered a cavern system beneath Montségur that led him to discover (apparently) a portion of this treasure. Otto Rahn was a fascinating person, an officer who was a scholar and a mystic. He believed in Atlantis, reincarnation, and that he had been a Cathar in a previous life.

According to the former US Army colonel Howard Buechner in *Emerald Cup—Ark of Gold*,[50] in February of 1944 the famous SS commando Otto Skorzeny recovered the treasure from the caverns beneath Montségur. Otto Rahn had died mysteriously the year before, apparently assassinated from within the Third Reich. According to Buechner, the treasure that was so carefully guarded for thousands of years eventually went to the Berchtesgaden fortress high in the Bavarian Alps, Germany's famed Eagle's Nest. In a scene worthy of the Indiana Jones film *Raiders of the Lost Ark*, the treasure was taken aboard a Nazi transport plane and eventually flown to a secret destination.

Says Buechner:

> In the very last days of April, 1945, eyewitnesses noted the mysterious takeoff, in the region of Salzburg, Austria, of a four engine aircraft believed to be a Heinkel 277 V-1 [Version-1]. The destination of the plane has remained unknown, but some authors have proposed that it flew to the city of Katmandu in Nepal and then to some other location in the Himalayas or in Tibet. On board were five officers of the Black Order and in the cargo bay were the twelve stone tablets of the Germanic Grail which contained the key to ultimate knowledge.[50]

Buechner believes that the Holy Grail and the Ark of the Covenant, as well as other treasures, were taken to a secret hideaway somewhere in Tibet. Buechner is apparently referring to German authors, writing in German, when he says, "some authors."

This may seem far-fetched, but there is a great deal of literature supporting close associations between the Third Reich and certain remote areas of Tibet whose locations are unknown. Given the Nazis' obsession with the occult, a destination such as Kathmandu or Tibet would not seem out of place. The suggestion that the flight might have landed in Kathmandu, rather than having flown directly to Tibet, is an interesting one. At this time in history, Nepal was a neutral nation, one that largely forbade outsiders from entering the country, but had strong ties to both Tibet and India, and many ethnic Tibetans live in Nepal. Most of Tibet's bronze, silver and gold statuary was actually made in the Kathmandu Valley and exported to Tibet.

The Nepalese probably would have allowed a German plane to land at their small airport on the fringes of Kathmandu, near the Himalayas, Tibet and Mount Everest. However, it would seem that the German planes might just as well have flown directly to the airbase in western Tibet, which is actually closer to Austria than Kathmandu.

Buechner suggests that the craft flying out of this far-eastern part of the Third Reich originated in Austria and flew directly east over Romania, the Black Sea and the Caspian Sea and into western Tibet. This flight can easily be made today by modern jets but during World War II, until its last year or so, such a flight was virtually impossible. Planes could carry drums of fuel on them and make stopovers along the way. Airports in Iran or Afghanistan, during the war, would be possibilities for such a stopover, but as we shall see, they were unnecessary.

There were long-range aircraft available at the end of the war and one of them was the Heinkel 277. The Heinkel He-277 was a four-engine, long-range heavy bomber design, originating as a derivative of the He-177. The He-177 used two Daimler-Benz DB 606 "power system" engines, while the He-277 used four similar engines.

The Heinkel 277 had a crew of seven and flew at a maximum speed of 570 km/h at 5,700 meters (354 mph at 18,700 feet) with a cruise speed of 460 km/h (286 mph). Its range was up to 11,100 km (6,900 miles) for the Amerika Bomber version that was to be able to bomb New York City and return without refueling. This plane was Heinkel's entry in the important transoceanic range

A Heinkel He 277 four-enigne bomber.

Amerikabomber competition with other aircraft manufacturers during the war to make long-distance flights, such as from France to New York City and back, without refueling in order to make long-range bombing missions against American targets.

However, it is said that the design was never produced and no prototype airframe was completed. Supposedly, the deteriorating condition of the German aviation industry late in the war and the competition from other long-range bomber designs from other firms, led to the design being cancelled. However, photos of completed He-277's are known, and shown here, and it is likely that at least four or five He-277s were made in the early process of converting the He-177 into a four-engine long-range bomber or cargo plane.

This was typical during World War II on all sides; the boys in the field did what they could and flew what they could, even if they were prototypes that did not fulfill all of the specifications of the final aircraft that was to the be the He-277. A number of these prototypes were undoubtedly built as the major components and the airframe basically already existed. The big question is where did these missing long-range planes fly to? Did one of them fly on a mission to Tibet at the end of WWII as Buechner and Barnhardt claimed? Also, it seems likely that the SS had early prototypes of the He-277 in 1942 or 1943. Apparently the early Haunebu II prototypes were available at this time.

The Heinkel-277 was one of four different long-range bomber prototypes that the Nazis had commissioned from the active aircraft manufacturing industry in Germany, Poland and the Czech areas. The He-277 had huge tires, long wings, four engines and a big glass dome area for the pilots and crew to sit in at the front of the fuselage. It would have been an impressive sight at any airfield in Europe or America at the time, and certainly would have

64

amazed people in remoter areas of the world, like Tibet, with its sheer size.

The Americans were doing much the same thing with the new aircraft that they had commissioned. If one industry was winning during WWII you could say it was the aerospace industry. Billions of dollars in wealth were poured into aerospace by all sides during this time. The Japanese were similarly developing their larger aircraft as well.

The Germans had four different companies creating large, long-range aircraft to continue the war on a larger scale. They were frustrated in their efforts to transfer technology to the Japanese via submarine and attempted flights over the pole to Japan, which were apparently successful.

Besides the He-277, there was the Messerschmitt Me-264, the Focke-Wulf Ta-400 and the Junkers Ju-390. All of these aircraft were long-range, four- or six-engine bombers (that could also be cargo planes), and the Third Reich pinned much of its effort to win the war on these new long-range strike aircraft. In fact, it was said by Albert Speer that a secret Ju-390 flight to Japan had occurred "late in the war." This flight, by a Luftwaffe test pilot, was a non-stop flight, probably from northern Norway, to Japan via the polar route. While some historians doubt that this flight took place, it probably was made and maybe more that once.

The Junkers Ju-390 carried a crew of ten with six engines and a longer range than the He-277; it made its maiden flight on October 20, 1943. It performed well and this resulted in an order for 26 aircraft, to be named Ju-390 V1.

Two prototypes were known to have been created by attaching an extra pair of inner-wing segments onto the wings of basic

A Messerschmitt Me 264 four-enigne bomber.

Junkers Ju-90 and Ju-290 airframes and adding new sections to lengthen the fuselages. It is thought that at least three Ju-390 prototypes were built and possibly many more. After the war it was claimed that the SS commandeered these prototype planes and put them at various SS bases in northern Norway, Greenland, and elsewhere—quite possibly Tibet.

On June 29, 1944, the Luftwaffe Quartermaster General paid Junkers to complete seven Ju-390 aircraft. The contracts for the 26 Ju-390 V1s were cancelled. Instead, a Version 2 (V2) was to be commissioned. At this point there were at least two prototypes of Ju-390 V1, and possibly as many as seven. They probably also had several other long-range prototypes, such as the He-277 prototypes, and therefore the SS at the very end of the war may have had 12 or more long-range aircraft that could carry passengers, cargo, machine parts and even arms to places as far away as Japan, South Africa and South America.

Atlantis, Tibet and Secret Science

Henry Stevens explains that the Nazis saw their origins in a Tibetan-Aryan culture of India, Tibet and the northern Baltic Regions. They also believed in Atlantis and in technical achievements made by civilizations in the remote past. Says Stevens:

> The problem was that European pre-history during the 1930s and before was centered about the Near East, so,

A Messerschmitt Me 264 four-enigne bomber in flight.

A Junkers Ju 390 six-engine bomber in flight.

allegedly, it was those people who were solely responsible for bringing "culture" to Europe. I say "Near East" but it took on tones of being bible-centric. The Nazis rejected this foreign origin of their culture and wanted to explore their own origins, both Indo-European and later Germanic.

The problem was that there were two "schools" to choose from in exploring native European origins. First there was the standard academic version, which, as I say, was certainly de-emphasized by mainstream archaeology and things would remain this way until the late 1960s. The other alternatives were the occult ideas of Madame H. P. Blavatsky, as reinterpreted by Guido von List and Lanz von Lebenfels, whose ideas found a home not in German science but in the Ahnenerbe, a Nazi organization which specifically sought Germanic-Aryan origins.

The Ahnenerbe became part of the SS in about 1940 and sent out expeditions of archaeologists and explorers in an attempt to establish a cultural and racial origin. Among these were the expeditions made to Tibet. Perhaps the Ahnenerbe thought the origins of the Germanics or the Aryans might be in Tibet, and so the grounds for measuring the populace in terms of typological anthropology, but this was not the view of the top scientists in German anthropology at the time, like Hans F.K. Guenther or Egon

67

Freiherr von Eickstedt. So please, let us not ever confuse the Ahnenerbe with German science other than the fact that the scientists employed by the Ahnenerbe were willing to be used for this purpose. This is certainly not unique to the Ahnenerbe—American anthropology and indeed all social science was and still is chock-full of such prostitutes who are directly or indirectly funded by the government through academia and dare not deviate from current political correctness [lest they] lose their jobs.[46]

But even if the Ahnenerbe entertained this racial-origin hypothesis on their first trip to Tibet, after that trip the reasons for continued expeditions must have been quite different. The Tibetans had asked for weapons and the Germans were probably happy to give them weapons. In exchange the Germans would get a secret airbase somewhere in Tibet.

Tibetan monks did make their way to Germany somehow and it was probably by aircraft. A letter was brought to Germany from the high Lama, and it seemed that Tibet had a pro-German stance at the outbreak of World War Two.

Stevens then theorizes that one of the things that the SS was interested in were the stories of the flying vehicles called vimanas in various Indian and Tibetan texts.[60] These ancient texts sometimes refer to the liquid metal mercury as being used in the vimanas. The Haunebu may have been propelled by "mercury gyros" that involved the electrification of a swirling mass of liquid mercury. The foo fighters might have used this same technology, as mentioned earlier. Stevens refers to the German writer Dr. Axel Stoll (*Hoch-Technologie im Dritten Reich*, 2001) who writes that the SS was interested in the various vimana texts:

Dr. Axel Stoll makes the point that the origins of some of the German flying disc technology may have been ancient India. Nobody really denies this is a possibility. Dr. Stoll, it seems, has really taken an interest in this possibility, translating the Vymaanika Shastra (evidently from an American/English version) into a scientific understanding in the German language. The point for us is that Dr. Stoll

 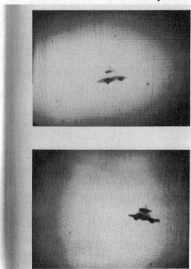

Four photos of a Haunebu in flight.

believes that it may be true that this knowledge survives and is guarded somewhere to this day, and suggests this place may be Tibet.

At a later point Dr. Stoll returns to Tibet in the book referenced above and describes it as a possible place for a secret German base. Here, he says outright that the Tibetans fought for Germany during the Second World War, so a good connection and rationale for such a base, at least in his mind, is evident. He says this would be an underground base and it would have been secured using electromagnetic means. This is what I have called the "electromagnetic vampire" in my book, *Hitler's Suppressed and Still-Secret Weapons, Science and Technology*. In chapter five of that book is a description of Dr. Stoll's electromagnetic security system. Suffice it to say that Dr. Stoll has his own unified field theory and in his interpretation, life force can simply be sucked out of an intruder using a sort of reverse-Faraday cage and the right electronics. In this underground secured base, among all the other things sequestered there, Dr. Stoll envisions flying discs.

So, the Nazis were interested in Tibet, possibly as a storehouse of ancient knowledge awaiting re-discovery through the efforts of the Ahnenerbe. We have evidence in the monks that people in some numbers were brought out of Tibet. It would have been almost impossible to smuggle them through India as Dr. Farrell points out. Dr. Stoll believes German underground base(s) would have been built, and that they would have contained, among other things, flying discs.[46]

So, now we not only have long-range aircraft based in Tibet but flying saucers as well. Stevens tells us that German flying discs, even the conventional ones, are credited by all German sources as being able to handle the Himalayas. He says the engineer Rudolf Lusar, who worked at the German Patent Office, gives a German flying disc the ability to climb, in three minutes, to an altitude of 12,400 meters or over 40,000 feet. Such a flying disk could hop over the Himalayas and Mt. Everest with ease. Flying from Tibet to anywhere in India after the war would not be a problem, since in the immediate postwar years the British had no real air defenses set up in India. Stevens thinks a secret base in Tibet would have housed flying saucers that were used to travel around Asia, as they were not really weapons. This secret Tibetan base may also be the source of the curious photos of flying saucers taken in China during WWII.

A Strange UFO Incident in Karachi

That this secret SS airbase was still active in Tibet in 1955 can be surmised by the curious UFO incident in Karachi, Pakistan, on March 23, 1955. The story of an obviously man-made flying disc comes from the late 90s British publication *UFO Magazine* in an article that is entitled "Incident at Karachi."

Pakistan had separated from India several years after the end of WWII in August of 1947 when the two countries gained self-rule from Great Britain. Karachi is located on the plains of Pakistan, to the southwest of Tibet. On that day in 1955 thousands of people in Karachi saw a huge flying disk near the airport. The disc was first observed by British military pilots and men who were returning from Singapore to Britain via Bangkok and Karachi. Many of these men were at 3,000 feet, aboard a DC-9 attempting to land at Karachi airport. Says Stevens about the report in *UFO Magazine*:

The teller of this tale is Mr. Frank J. Parker, who in the late 1990s was living in Hyde, Cheshire, England. Mr. Parker, twenty years old at

A Haunebu entering a cloud.

the time, was onboard the DC-9 in question and so first observed the flying saucer after some excited talk in the passenger cabin. Because the passengers from one side of the aircraft were standing up and moving to the other side of the aircraft, the captain ordered all the passengers back to their seats to avoid imbalance of the aircraft. Well, I have flown on a DC-6. It was a large, four- engine passenger aircraft. If it was in danger of becoming imbalanced, there were certainly many individuals moving from one side to the other observing this flying craft, and so it must have caused great excitement.

Upon landing, the military men boarded a bus for the 16- or 17-mile drive to Karachi. They could still see the flying saucer, stationary and suspended over the city. At first, according to Parker, from the air it looked like an orange ball, but took on a disc shape as they drew nearer to it. As they continued on toward it, Parker estimated that it was 300-400 feet in the air. The bus turned a corner on to "the parade grounds" where Parker describes the sight that met his eyes as "incredible." Thousands of people, mostly Pakistanis, were amassed on the parade grounds, and above them, as if suspended, was the flying disc. Many of the Pakistanis knelt on the ground, hands clasped as though they were praying. Parker and some companions exited the bus and began moving through the crowd towards the spot over which the flying disc hung. Nobody was afraid or upset, according to Parker, and so the British party was

71

able to move through the kneeling crowd without taking their eyes off the flying disc. They came to a halt directly below the still-suspended disc, now only about 60 yards above them.[46]

It is interesting to note here that the Pakistanis, all Muslims, were kneeling on the ground and praying. For them it was a sign from Allah. How many other historic incidents have been created by a brightly lit circular craft coming down from the sky?

Frank Parker then describes the scene before them:

This had to have been at least 150 feet across and two-to-three stories high—by that I mean about 75 feet, it was incredible. Because it was hovering about 12-feet off the ground, we could see every detail. This was a 'nuts and bolts' spacecraft that looked man-made, because we could see what I could only describe as rivets, only perfectly aligned.

I suppose because of its shape you could say it was a traditional 'flying saucer,' like two bronze cymbals, one on top of the other. It was magnificent, made up entirely of this metallic bronze-colored metal that applied to the catwalks, ladders, portholes and aerials. These were all fixed to the exterior of the craft. At the underside, we saw a big black hole, which interested me because whatever was holding this thing up had to come from there. Well, we saw what looked like fans and I counted eight rotors, but every few seconds there would be a 'whooshing' sound and that created sparks and an array of laser-like effects.

With catwalks, ladders, portholes, aerials, and rotors generating a downdraft, we can pretty much say that this craft was not some interstellar space-craft. Parker continues, discussing the "whoosh-ing" sound:

Whenever this occurred, the entire craft would swing to

the left and right and then come back again. This happened so many times that I believed whoever was behind the controls was experiencing engine problems... Every now and again, the craft ejected smoke towards the ground, and this gave off a strong smell of oil, but here's another curious thing—this stuff actually never reached the ground—rather it just seemed to evaporate... The craft made a constant noise not dissimilar to jet-engine propulsion, but nothing like as loud. This and the noise of the rotors would sometimes sound rough, as though they were faltering occasionally. Anyway, every now and again, it would force hot air downwards and this had the effect of blowing dust and pieces of paper all over the place.[46]

This fascinating incident would appear to involve one of the German flying disks coming from Tibet, perhaps a Haunebu II. What it was doing in the area is difficult to determine. It would seem to have been on a mission to simply harass the British forces in Karachi—in 1955—a group that mainly serviced transport aircraft carrying soldiers back to Britain from India and the Far East. Perhaps the craft was there to pick someone up or let people off in a park where they could just melt into the citizenry of the city. From Karachi one could easily get trains, ships and airplanes to other parts of the world.

Once again, western Tibet would be a good geographical spot for this secret airbase to exist, and the Chang Tang Highlands of the north section of western Tibet is an area that is extremely remote, even today, with few roads or towns. It should be noted here that except for water, the Germans would have needed very little from the surrounding terrain. The more remote, the better (as we see with the bases in Antarctica and elsewhere). Like these bases, the base in Tibet would have been supplied from the air, with aircraft—and incredibly, Haunebu flying saucers—bringing machinery, operators, parts and food to these secret locations.

The secret Tibet base was just one of many secret bases that the SS had set up during the war. As I chronicle in in my book *Antarctica and the Secret Space Program*, the Germans had set up a submarine base in Neuschwabenland in 1938. They went on to create other secret bases, often for submarines, in Norway, Greenland, The Canary Islands, the Spanish Sahara and other areas. These so-called secret cities of the Black Sun will be discussed at greater length in another chapter.

The Haunebu photo negative given to Henry Stevens by Vladimir Terzinsky.

Chapter 2

World War II as an Oil War

Did you exchange
A walk on part in the war
For a lead role in a cage?
—*Wish You Were Here*, Pink Floyd

One of the aspects of WWII that is often neglected is that much of the war was a fight over oil fields and oil facilities. The struggle for energy reserves by the Axis powers was a major part of their strategy—and failure—during the war. Let us examine the oil war aspect of WWII and the ultimate rise of the Haunebu, Vril and other electric craft.

First of all, we need to explain that the main source of oil for the Nazis was the oilfields in Romania and their synthetic oil plants in Germany—a country with virtually no oilfields of its own. Romania was an early oil-exporting nation starting in the 1800s. It was also the largest oil exporting European country during WWII and all of its exported oil went to the Third Reich and its occupied territories. Romania was not an occupied country during the war but was a Kingdom that readily joined the Axis powers and participated in the invasion of the Ukraine and Russia. As the Third Reich expanded, its need for oil also expanded and this created a tremendous problem for the Nazi regime: they needed energy for their massive war machine. Typically, this energy was in the form of oil derivatives such as gasoline, aviation fuel and such, much of it coming from several large oil installations around Ploiesti, Romania. In Germany itself were several synthetic fuel plants.

These oilfields were vulnerable and were attacked later in the war. But even earlier, the oil produced at Ploiesti, and from other sources, was not enough oil for the Nazi war machine. They had lots of tanks, armored cars, troop carriers and airplanes. What they didn't have was lots of fuel. Therefore, historians now realize that WWII was essentially an oil war. The Germans launched Operation Barbarossa, the invasion of the Soviet Union for two reasons: 1) to gain territory and "living space" (Lebensraum) and 2) to seize the large oilfields at Baku on the Caspian Sea in what is today Azerbaijan. Germany, which wanted to expand it territory, needed oilfields. The ones they set their sights on were the massive oilfields at Baku, oilfields that had already been fought over in WWI.

A special study of WWII as an oil war was done by Joel Hayward and published in the *Journal of Strategic Studies* in 1995. In his lengthy article titled "Hitler's Quest for Oil: The Impact of Economic Considerations on Military Strategy" Hayward says at the beginning:

> When asked by his Allied captors in 1945 to what extent German military strategy had been influenced at various stages by economic considerations, Albert Speer, Hitler's outstanding Armaments Minister, replied that in the case of Operation Barbarossa the need for oil was certainly a prime motive. Indeed, even during the initial discussions of his plan to invade the Soviet Union, Hitler stressed the absolute necessity of seizing key oilfields, particularly those in the Caucasus region, which accounted for around 90 per cent of all oil produced in the Soviet Union. For example, during a war conference at the Berghof on 31 July 1940, Hitler revealed to high-ranking commanders his intention to shatter Russia 'to its roots with one blow.' After achieving the 'destruction of Russian manpower,' he explained, the German Army must drive on towards the Baku oilfield, by far the richest of those in the Caucasus and one of the most productive in the world.

Operation Barbarossa and the Oil War

Operation Barbarossa (German: *Unternehmen Barbarossa*) was the code name for the Axis invasion of the Soviet Union, which started on June 22, 1941. Operation Barbarossa put into action Nazi Germany's ideological goal of conquering the western Soviet Union so as to repopulate it with Germans, as per the idea of Lebensraum (living space). The German *Generalplan Ost* aimed to use some of the conquered people as slave labor for the Axis war effort while acquiring the oil reserves of the Caucasus as well as the agricultural resources of various Soviet territories. Their ultimate goal included the eventual extermination, enslavement, Germanization and mass deportation to Siberia of the Slavic peoples, and to create more Lebensraum for Germany. The plan was to take land as far east as Moscow and as far south as Armenia and the oilfields at Baku. Says Wikipedia of Operation Barbarossa and the heavy cost of life over the next four years:

> In the two years leading up to the invasion, Germany and the Soviet Union signed political and economic pacts for strategic purposes. Nevertheless, the German High Command began planning an invasion of the Soviet Union in July 1940 (under the codename Operation Otto), which Adolf Hitler authorized on 18 December 1940. Over the course of the operation, about three million personnel of the Axis powers—the largest invasion force in the history of warfare—invaded the western Soviet Union along a 2,900-kilometer (1,800 mi) front, with 600,000 motor vehicles and over 600,000 horses for non-combat operations. The offensive marked an escalation of World War II, both geographically and in the formation of the Allied coalition including the Soviet Union.
>
> The operation opened up the Eastern Front, in which more forces were committed than in any other theater of war in history. The area saw some of the world's largest battles, most horrific atrocities, and highest casualties (for Soviet and Axis forces alike), all of which influenced the course of World War II and the subsequent history of the 20th century. The German armies eventually captured some five million Soviet Red Army troops, a majority of whom never returned alive. The Nazis deliberately starved to death, or otherwise killed, 3.3 million Soviet prisoners of

war, and a vast number of civilians, as the "Hunger Plan" worked to solve German food shortages and exterminate the Slavic population through starvation. Mass shootings and gassing operations, carried out by the Nazis or willing collaborators, murdered over a million Soviet Jews as part of the Holocaust.

As early as 1925, Adolf Hitler vaguely declared in *Mein Kampf* that he would invade the Soviet Union, asserting that the German people needed to secure Lebensraum ("living space") to ensure the survival of Germany for generations to come. On 10 February 1939, Hitler told his army commanders that the next war would be "purely a war of Weltanschauungen ["worldview"] ...totally a people's war, a racial war." On 23 November, once World War II had already started, Hitler declared that "racial war has broken out and this war shall determine who shall govern Europe, and with it, the world." The racial policy of Nazi Germany portrayed the Soviet Union (and all of Eastern Europe) as populated by non-Aryan Untermenschen ("sub-humans"), ruled by Jewish Bolshevik conspirators.

A map of Greater Germany with Baku in the bottom right corner.

The Concept of Lebensraum or "Living Space"

The main reasons for the invasion of the Soviet Union stemmed from the concept of Lebensraum and the quest for oil and other resources. The German ethnographer and geographer Friedrich Ratzel in his 1897 book *Politische Geographie* applied the word Lebensraum to describe physical geography as a factor that influences human activities in developing into a society. Ratzel extended his thesis in his 1901 essay titled "Lebensraum." Other philosophers and politicians picked up on the concept during World War I. At that time, when the British blockade of trade coming into Germany caused food shortages, these food shortages caused support to rise for a German Lebensraum expansion that would extend Germany eastward into Russia to gain control of more farmland and other resources to stop the chronic deprivations.

During the period of 1919 to 1939, between the First and the Second World Wars, German nationalists adopted the term Lebensraum to refer to their plan for the establishment of a Germanic colonial empire like the British Empire, the French Empire, and the empire that the U.S. established by going westward, which was justified by the ideology of Manifest Destiny first espoused in 1845. Adolf Hitler wrote about Lebensraum in his book *Mein Kampf,* and had an entire chapter on how Germany needed to expand eastward into Russia and the Caucasus Mountains. The geopolitician, occultist and Vril Society founding member Karl Haushofer provided the Nazis with intellectual, academic, and scientific rationalizations for Lebensraum, transmitted to Chancellor Adolf Hitler by way of Rudolf Hess, who was Haushofer's student. Says Wikipedia:

Prof. Friedrich Ratzel.

With Adolf Hitler's rise to power, Lebensraum became an ideological principle of Nazism and provided justification for the German territorial expansion into Central and Eastern Europe. The Nazi *Generalplan Ost* policy ('Master Plan for the East') was based on its tenets. It stipulated that Germany required a Lebensraum necessary for its survival and that most of the indigenous populations of Central and Eastern Europe would have to be removed permanently (either through mass deportation to Siberia, extermination, or enslavement) including Polish, Ukrainian, Russian, Czech and other Slavic nations considered non-Aryan. The Nazi government aimed at repopulating these lands with Germanic colonists in the name of Lebensraum during World War II and thereafter. Entire indigenous populations

Adolf Hitler with Field Marshal von Brauchitsch in 1941.

were decimated by starvation, allowing for their own agricultural surplus to feed Germany.

Hitler's strategic program for world domination was based on the belief in the power of Lebensraum, especially when pursued by a racially superior society. People deemed to be part of non-Aryan races, within the territory of Lebensraum expansion, were subjected to expulsion or destruction.

Hitler was warned by his general staff that occupying "Western Russia" would be costly and create "more of a drain than a relief for Germany's economic situation." However, Hitler anticipated compensatory benefits, such as the exploitation of the Ukraine as a reliable and immense source of agricultural products; the use of forced labor to stimulate Germany's overall economy; control of the oil fields at Baku; and the expansion of territory to help isolate the United Kingdom, including occupying ports on the Black Sea. Hitler was convinced that Britain would sue for peace once the Germans triumphed in the Soviet Union, and if they did not, he would use the resources available in the east such as the extra fuel from Baku to defeat the British Empire.

> *We only have to kick in the door and*
> *the whole rotten structure will come crashing down.*
> —Adolf Hitler on the Soviet Union

The invasion plans were drawn up by the German military elite, who had just seen the rapid defeat of France at the hands of the "invincible" Wehrmacht, and were confident of their defeat of Russia, which they viewed as a primitive, backward "Asiatic" country. Red Army soldiers were considered brave and tough, but their officers were held in contempt. Hitler received the final military plans for the invasion on which the German High Command had been working since July 1940 under the codename "Operation Otto." Hitler, dissatisfied with these plans, issued Führer Directive 21, which called for a new battle plan, now code-named "Operation Barbarossa." The operation was named after medieval Emperor Frederick Barbarossa of the Holy

Roman Empire, a leader of the Third Crusade in the 12th century. Hitler approved the plans for Operation Barbarossa on December 5, 1940. The invasion was set for 15 May 1941, though it was delayed for over a month to allow for further preparations and possibly better weather.

Hitler and the General Staff did not anticipate a long campaign that would last into the winter, therefore preparations such as the distribution of warm clothing and the winterization of vehicles were not made.

In 1941, Nazi ideologue Alfred Rosenberg—later appointed

General Ewald von Kleist (left), commander of the 1st Panzer Group, inspects a large iron works facility in Ukraine, 1941

Himmler inspects a prisoner of war camp in Russia, circa 1942.

Reich Minister of the Occupied Eastern Territories—suggested that conquered Soviet territory should be administered in the following Reichskommissariate ("Reich Commissionerships"). These were five administrative subdivisions of conquered Soviet territories:

1. Reichskommissariat Ostland: Baltic countries and Belarus (realized 1942)
2. Reichskommissariat Ukraine: Ukraine, enlarged eastwards to the Volga (realized 1942)
3. Reichskommissariat Kaukasus: Southern Russia and the Caucasus region (Unrealized)
4. Reichskommissariat Moskowien: Moscow metropolitan area and remaining European Russia (Unrealized)
5. Reichskommissariat Turkestan: Central Asian republics and territories (Unrealized)

83

The Reichskommissariat Kaukasus was perhaps the most important of the Reichskommissariats as it held the oilfields of Baku. Occupying the Ukraine would have to occur before the German Army could move south to the Caucasus area. This operation to take the Caucasus area and the oilfields at Baku was called Operation Edelweiss, a separate operation within Operation Barbarossa.

The Battle of the Caucasus and Operation Edelweiss

The Battle of the Caucasus is a name given to a series of Axis and Soviet operations in the Caucasus area on the Eastern Front. On July 25 1942, German troops captured Rostov-on-Don, Russia, opening the Caucasus region of the southern Soviet Union, and the oil fields beyond at Maikop, Grozny, and ultimately Baku, to the Third Reich. Two days earlier, Adolf Hitler had issued a directive to launch the assault into the Caucasus region, Operation Edelweiss.

Operation Edelweiss, named after the mountain flower, aimed to gain control over the Caucasus and capture the oil fields of Baku. The main forces included Army Group A, comprised of 1st Panzer Army, 4th Panzer Army, 17th Army, part of the Luftflotte 4 and the 3rd Romanian Army. This Army Group A was supported to the east by Army Group B and by the remaining 4th Air Fleet aircraft (1,000 aircraft in all). In anticipation of taking the oilfields at

German soldiers in the Caucasus Mountains, 1942.

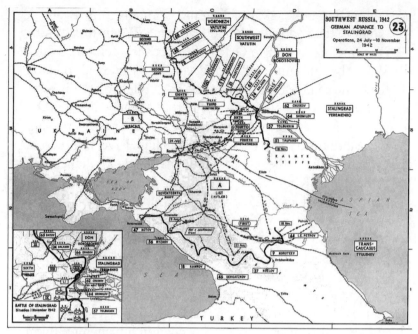

A map of the front in the Cacausus area from July to November 1942.

Baku, these land forces were accompanied by 15,000 oil industry workers and included 167,000 troopers, 4,540 cannons, and 1,130 tanks. Says Wikipedia:

> Several oil firms such as "German Oil on Caucasus," "Ost-Öl" and "Karpaten-Öl" had been established in Germany. They were awarded an exclusive 99-year lease to exploit the Caucasian oil fields. For this purpose, a large number of pipes—which later proved useful to Soviet oil industry workers—were delivered. A special economic inspection "A," headed by Lieutenant-General Nidenfuhr was created. Bombing of the oil fields was forbidden. To defend them from destruction by Soviet units under the command of Nikolai Baibakov and Semyon Budyonny, an SS guard regiment and a Cossack regiment were formed. The head of the Abwehr developed Operation Schamil, which called for landing in the Grozny, Malgobek and Maikop regions. They would be supported by the local fifth column.
>
> After neutralizing the Soviet counter-attack in the

Izyum-Barvenkovsk direction the German Army Group A rapidly attacked towards the Caucasus. When Rostov-on-Don, nicknamed "The Gates of Caucasus," fell on 23 July 1942, the tank units of Ewald von Kleist moved towards the Caucasian Mountain Range. The "Edelweiss" division commander, Hubert Lanz, decided to advance through the gorges of rivers of the Kuban River basin and by crossing the Marukhskiy Pass (Maly Zelenchuk River), Teberda, Uchkulan reach the Klukhorskiy Pass, and simultaneously through the Khotyu-tau Pass block the upper reaches of the Baksan River and the Donguz-Orun and Becho passes.

Concurrently with the outflanking maneuvers, the Caucasian Mountain Range was supposed to be crossed through such passes as Sancharo, Klukhorskiy and Marukhskiy to reach Kutaisi, Zugdidi, Sukhumi and the Soviet Georgian capital city of Tbilisi. The units of the 4th German Mountain Division, manned with Tyroleans, were active in this thrust. They succeeded in advancing 30 km toward Sukhumi. To attack from the Kuban region, capture the passes that led to Elbrus, and cover the "Edelweiss" flank, a vanguard detachment of 150 men commanded by Captain (Hauptmann) Heinz Groth, was formed. From the

German tanks in formation in a Caucasus valley with infantry in the foreground, September 1942.

Old Karachay through the Khurzuk aul and the Ullu-kam Gorge the detachment reached the Khotyu-tau Pass, which had not been defended by the Soviet troops. Khotyu-tau gained a new name – "The Pass of General Konrad."

The starting point of the operation on the Krasnodar-Pyatigorsk-Maikop line was reached on 10 August 1942. On 16 August the battalion commanded by von Hirschfeld made a feint and reached the Kadar Gorge. On 21 August troops from the 1st Mountain Division planted the flag of Nazi Germany on the summit of Mount Elbrus, the highest peak in both the Caucasus and Europe.

The German and Romanian troops made fairly good headway through the Caucasus region in 1942, taking Stravropol on August 3, and the town of Krasnodar on August 12 and the town of Mozdok by August 25. The blitzkrieg stopped at the Chechen-Ingush ASSR town of Malgobek and the North Ossetia city of Vladikavkaz in September of 1942. This was the furthest south that the Axis powers would get, only a few hundred kilometers to the oilfields of Baku. They did not make any more progress and ultimately began to retreat in January of 1943. After the Soviet breakthroughs in the region around Stalingrad, the German forces in the Caucasus were put on the defensive. Red Army units retook the city of Mozdok in January 1943, forcing German forces to begin a slow retreat.

Because of setbacks elsewhere on the German front—not at the Caucasus front—in early 1943 the Germans began to withdraw and consolidate their positions. The German Army established a defensive line at the Kuban bridgehead in the Taman Peninsula and from here they hoped to launch new operations in the Caucasus during the summer. The fighting remained reasonably static until September of 1943 when the Germans ordered fresh withdrawals and this effectively ended the Battle of the Caucasus and the quest for the oilfields at Baku.

The Destruction of the Romanian Oilfields

The failure of Operation Barbarossa reversed the fortunes of the Third Reich. The Axis powers never got the needed petroleum facilities at Baku and they now had a shortage of fuel, especially on the Eastern Front. Operationally, German forces had achieved significant victories and occupied some of the most important economic areas of the Soviet Union (mainly in Ukraine). The Germans inflicted heavy casualties on the Soviets and sustained heavy casualties themselves. But despite their early successes, the German offensive stalled in the Battle of Moscow at the end of 1941, and the subsequent Soviet winter counteroffensive pushed German troops back. The Germans had been confident in a quick collapse of Soviet resistance, but the Soviet Army absorbed the German Wehrmacht's strongest blows and bogged it down in a war of attrition for which the Germans were unprepared. The Germans undertook several operations to retake the initiative and drive deep into Soviet territory such as Case Blue in 1942 and Operation Citadel in 1943. These operations eventually failed, which resulted in the Wehrmacht's retreat and the collapse of the Eastern Front.

But things would get worse for the Germans. The oilfields in the vicinity of Ploiesti were to come under attack in June of 1942 and again in August of 1943. As I mentioned earlier, the Ploiesti oil facilities in Romania were a major source of fuel for the Nazi war machine. Romania was the largest producer of oil in Europe

A petroleum field at Moreni, Romania, 1920.

during WWII. The Mehedințeanu brothers opened the world's first large refinery at Ploiesti in 1856–1857. History also remembers the city as the site of the self-styled Republic of Ploiesti, a short-lived 1870 revolt against the Romanian monarchy.

The first attack on the Ploiesti oil facilities was carried out by the Halverson Project or HALPRO, a group of 23 B-24 Liberator heavy bombers under the command of Col. Harry Halverson.

The U.S. Army Air Forces (AAF), the predecessor to the U.S. Air Force, prepared for global war in January 1942. With the enemy in Europe and Asia, Allied forces would have to span the globe, and rely heavily on air power if they wanted to be successful. One plan focused on establishing American air forces in the China-Burma-India Theater of Operations. Light bombers, cargo planes, and fighters would establish the Tenth Air Force, which would include a group of 23 B-24 Liberator heavy bombers, commanded by Col. Harry Halverson. With handpicked crews, this group became known as the Halverson Project or HALPRO. Their planned assignment included raids on the Japanese Home Islands.

Halverson, who became a pilot during World War I, had vast experience in the air. In 1924 he served as ground support organizer for the Army's Round the World Flight in India. Because of his decades of aviation experience, Halverson was an obvious choice to command the long-range bombers.

The HALPRO group departed the United States on May 22, 1942 flying to West Africa and then to Khartoum, Sudan. It then flew on to a British airbase in northeast Egypt. Having flown around 4,000 miles, maintenance problems grounded some of the arriving aircraft. This would be the first AAF attack over Europe, and Halverson maintained secrecy and independence of command.

On June 11 at 10:30 p.m., the 13 available B-24 Liberator heavy bombers took off for Romania. Halverson directed crews to fly directly to Romania, and return to Habbaniyah, Iraq, even though this route violated neutral Turkish airspace. One plane bombed the Romanian harbor of Constanta and turned back. The other 12 Liberators approached Ploiesti at 14,000 feet. Unfortunately bad weather frustrated navigators, making the bombing inaccurate and creating only light damage, if any. The raid was considered

89

A B-24 Liberator during Operation Tidal Wave in 1943.

a failure.

The crews dropped 24 tons of bombs around Ploesti most of them doing little damage while antiaircraft fire and a few enemy fighters were active against the group. Halverson's aircraft and three others flew to Iraq and landed at Habbaniyah as planned, while five landed at other British-controlled airfields in Iraq and Syria, and four landed in neutral Turkey and were then interned. With little damage inflicted and some crews interned, Allied leadership considered the raid a failure and Gen. Dwight D. Eisenhower remarked dryly that the raid "did something to dispel the illusion that big planes could win the war."

Operation Tidal Wave

One year later in the summer of 1943 the oilfields were attacked again in what was called Operation Tidal Wave. The earlier bombing caused minimal damage, and if anything the facilities were now putting out more petroleum products, rather than less. Germany and Romania responded to the 1942 attack by putting strong anti-aircraft defenses around Ploiesti. Luftwaffe General Alfred Gerstenberg built one of the heaviest and best-integrated air defense networks there during the war. The defenses included several hundred large-caliber 88 mm guns and 10.5 cm Flak 38

B-24 Liberators at low altitude while approaching the oil refineries at Ploiesti, Romania, August 1, 1943. In the foreground is B-24D *Li'l Jughaid*, serial number 42-63758, Ninth Air Force, 98th Bombardment Group, 415th Bombardment Squadron. Pilot Lyle Spencer, co-pilot Harry J. Baker, bombardier Boyden Supiani. Photograph was taken from the 98th's lead plane, piloted by Col. John R. Kane. Following in formation behind *Li'l Jughaid* are *Daisy Mae* and *Black Magic*.

91

anti-aircraft guns, and many more small-caliber guns. Many of these guns were concealed in haystacks, railroad cars, and mock buildings. The Ploiesti oilfields were the most heavily defended Axis target outside the Third Reich.

Operation Tidal Wave was to comprise of 178 B-24 Liberator bombers of the United States Army Air Forces (USAAF) based in Libya. They were to attack nine oil refineries around Ploiesti on August 1, 1943. It was the most expensive and important of the strategic bombing missions flown as part of the "oil campaign" to deny petroleum-based fuel to the Axis powers.

The planes were to fly from British airfields near Benghazi, Libya north across the Mediterranean and the Adriatic Sea, pass near the island of Corfu, cross over the Pindus Mountains in Albania, cross southern Yugoslavia, enter southwestern Romania, and turn east toward Ploiesti. They were to come into Ploiesti from the north where they were to locate pre-determined checkpoints and strike all targets simultaneously. They had been told to avoid the city of Ploiesti and not bomb it by accident. They were only to bomb the oil facilities, which were scattered in the countryside north of the city. After they dropped their bombs they were to continue flying south back to the airfields in Libya, or alternately, to airfields in Cyprus.

Early on the morning of August 1 1943, the five groups comprising the strike force of 178 bombers began lifting off from their home airfields around Benghazi. Large amounts of dust kicked up during takeoff caused limited visibility and strained engines already carrying the burden of large bomb loads and additional fuel. These conditions contributed to the loss of one aircraft during takeoff.

The formation of now 177 B-24 Liberators reached the Adriatic Sea without further incident; however aircraft #28, *Wongo Wongo*, belonging to the 376th Bombardment Group (the lead group, about 40 B-24s) began to fly erratically before plunging into the sea due to unknown causes. Aircraft #23 *Desert Lilly*—descended from the formation in order to look for survivors, narrowly missing aircraft *Brewery Wagon*. No survivors were seen, and due to the additional weight of fuel, the pilot was unable to regain altitude to rejoin the formation and resume course to Ploiesti.

The group had been told to maintain strict radio silence and therefore there was confusion among this group as to what was happening, and ten other aircrews returned to friendly airfields after the incident. The remaining aircraft then made the 9,000-foot (2,700 meter) climb over the Pindus Mountains of Albania, which were shrouded in cloud cover.

The five bomber groups were told to fly low over Romania to avoid German radar. At the town of Câmpina the group split up as previously planned, however some of the bombers began following the wrong railway tracks, ones that would not lead them to the refineries as planned. When other bombers noticed this they broke radio silence; at the same time the many air defense batteries around Câmpina and Ploiesti began to fire at them and fighter planes were scrambled to intercept them.

The Columbia Aquila refinery in Romania burning after the raid of B-24 Liberator bombers, Operation Tidal Wave. The refinery was largely intact.

A map of the approximate flight route for Operation Tidal Wave.

All hell was about to break loose on the ground and in the air. The *Hell's Wench* B-24, flown by Lt. Col. Baker and his co-pilot Maj. John Jerstad, was now hit heavily by flak. They jettisoned their bombs to maintain the lead position of the formation over their target at the Columbia Aquila refinery. Despite heavy losses by the 93rd, Baker and Jerstad maintained course and once they were clear of the refinery began to climb upward away from the formation flying behind them. Realizing that their aircraft was no longer controllable, the pilots kept climbing to let their crew abandon the aircraft. None of the crew survived and Baker and Jerstad were posthumously awarded the Medal of Honor for these actions.

Meanwhile, Maj. Ramsay D. Potts flying *The Duchess* and Maj. George S. Brown flying the *Queenie*, encountering heavy smoke over the Columbia Aquila oil facility, led additional aircraft of the 93rd squadron that successfully dropped their bombs over the Astra Romana, Unirea Orion, and Columbia Aquila refineries. In the end, the 93rd lost 11 aircraft over their targets around Ploiesti. One of these bombers, the *Jose Carioca*, was shot down by a Romanian IAR 80 fighter, which went into a half roll and moved swiftly under the B-24 upside down, raking its belly with bullets and killing some of the crew. The bomber then crashed into the Ploiesti Women's Prison killing all aboard. This three-story building exploded in flames, killing over 50 people and wounding hundreds of others.

Over the Romana American refinery the 376th Bomb Group encountered heavy air defense fire and portions of this group went to the Steaua Română refinery at Câmpina while others headed for the smoldering conflagration over the Concordia Vega refinery. At the town of Câmpina where a number of refineries were located, air defenses on overlooking hills were able to fire down into the formation of B-24s.

Meanwhile, the 98th Bomb Group and the 44th Bomb Group made their prescribed turn at Floresti and proceeded to their respective targets at the Astra Romana and Columbia Aquila refineries. Both groups found German and Romanian defenses on full alert and faced the full effects of now raging oil fires, heavy smoke, secondary explosions, and delayed-fuse bombs dropped by the 93rd Bomb Group on their earlier run. These bombers approached at low altitude flying parallel to the Floresti to Ploiesti railway and suddenly encountered a disguised flak train.

Flying at treetop level the bombers were now being fired upon by the guns on this moving train. However, the gunners on the B-24s quickly responded to the threat, disabling the locomotive and killing multiple air defense crews on the disguised train.

A Cartwheel Bomber Crash and the Return to Libya

The last Operation Tidal Wave attack bombed the Steaua Română refinery at Câmpina. The 389th and 376th bomber groups led the attack that heavily damaged the refinery, which did not resume production for the duration of the war. The 389th lost four aircraft over the target area, including B-24 *Ole Kickapoo* flown by 2nd Lt. Lloyd Herbert Hughes. After hits to *Ole Kickapoo* only 30 feet over the target area, the detonation of previously dropped bombs had ignited fuel leaking from the B-24. Hughes maintained course for his bombardier to bomb the refinery, and the B-24 subsequently crash-landed in an explosive cartwheel into a riverbed. Hughes (who posthumously received the Medal of Honor) and five crewmembers were killed, but incredibly four survived the crash but died of injuries, and two gunners became prisoners of war.

Only 88 B-24 Liberators returned to Libya, of which 55 had battle damage. Losses included 44 to air defenses and additional B-24s that ditched in the Mediterranean or were interned after landing in neutral Turkey. Some were diverted to the RAF airfield on Cyprus. One B-24 landed in Libya 14 hours after departing with 365 bullet holes in it; its survival was due to the light armament of the Bulgarian Avia B-534 fighter's machine guns that had attacked it on its return to Libya.

Of the hundreds of American crewmembers, 310 airmen were killed, 108 were captured by the Axis, 78 were interned in neutral Turkey, and four were taken in by Tito's partisans in Yugoslavia. Three of the five Medals of Honor (the most for any single air action in history) were awarded posthumously.

The raid was considered an overall failure. Although the Allies estimated a loss of 40% of the refining capacity at the Ploiesti refineries, some refineries were largely untouched. Sadly, most of the damage was repaired within weeks, after which the net output of fuel was greater than before the raid. Many of the refineries had been operating previously below maximum capacity and now were repaired and producing more fuel than ever.

The oil war with Nazi Germany would continue, with the Allies unable to stem the flow of oil from Romania. On the other hand, the Axis powers had failed to take the oil fields at Baku, one of their most important goals. It was sort of a score of one to one in the WWII oil wars.

The Operation Tidal Wave mission resulted in "no curtailment of overall product output." It was one of the costliest missions for the USAAF in the European Theater. It was proportionally the most costly major Allied air raid of the war, and its date was later referred to as "Black Sunday." A research report prepared in 1999 for the Air War College at Maxwell Air Force Base in Alabama concluded that the mission to Ploiesti was "one of the bloodiest and most heroic missions of all time."

Soviet troops captured Ploiesti in August of 1944 and that was the end of the royal family. Following the war, the new Communist régime of Romania nationalized the oil industry, until then privately owned, and made massive investments in the oil and petroleum industry in a bid to modernize the country and to repair the war damage. Romania is still a major oil producer.

So, we see how WWII was really an oil war, though this is rarely mentioned in documentaries of WWII. Similarly, Japan did not have its own oil resources and took oil fields in Manchuria. Italy maintained oilfields in Libya until they were taken from them by the British. War machines need fuel and the Axis struggled to gain strategic oilfields but failed. Now alternative technologies pioneered by Nikola Tesla and others would come into play. Were there more exotic technologies that were electric in nature? Indeed, there were. While the traditional military wanted to rely on their tanks, troop carriers and proven aircraft, others in the SS turned to more exotic aircraft, including flying disks.

It is possible that a Haunebu or Vril craft took part in the interception of the bombers of Operation Tidal Wave? We are told that aircraft #28, *Wongo Wongo*, belonging to the lead 376th Bombardment Group, began to fly erratically before plunging into the Adriatic Sea due to unknown causes. Because of the radio silence he would not communicate what had happened to the plane. They had clearly lost control of their aircraft. Since this was the lead aircraft, had a Haunebu II come from above the squadron and used an electromagnetic ray to disable the *Wongo Wongo*?

Though the chances are slim, we do not know what caused the bomber to crash, but the Haunebu II seems to have been deployed about this time as a scout craft. But more important is that the WWII oil war had the effect of luring the Third Reich to the Caucasus region to get the oilfields at Baku and their failure there doomed the dream of a Greater Germany.

As the failures on the Russian Front began to mount and the Nazis were forced to retreat, things became more desperate within the Third Reich. Hitler had angrily denounced his generals for allowing the retreat and wanted them to fight to the last man.

Attempts to assassinate Hitler were made but the Third Reich continued to be on the defensive, and despite the oil output from Romania were having a severe shortage of petroleum products.

The special projects like the Bell, the Haunebu, and latest U-boat technology were the areas in which to focus their technological research. Ultimately the surviving U-boats, long-range aircraft and the Haunebu were used to transport people, documents and gold out of the failing Reich to the secret bases that had been set up even before the war. Ultimately these surviving SS officers and high officials would head to South America, some via the secret submarine base in Antarctica and some by a journey in a Haunebu.

Chapter 3

The Volga Germans

It's a riddle, wrapped in a mystery,
inside an enigma, but perhaps there is a key.
—Winston Churchill, BBC address, October 1939

One group of people caught up in the oil war of WWII were the Volga Germans. Their story is a curious and tragic one, and a story that involves the Haunebu. The Volga Germans were from Germany, lived in the Soviet Union, and in colonies in Argentina. Their role in the flights of the Haunebu, Vril and Andromeda craft must have been critical, as we shall see.

Part of Germany's plan to hold the Caucasus oilfields and other parts of the western Soviet Union was that a fifth column of partisans and revolutionaries would rise up against their Soviet oppressors and be happy to work with the Nazis. One such group was the Volga Germans, a large population of Germans that were already living the Soviet Union and had their own small state within the USSR. Their role in WWII is rarely talked about and surviving members of this group are linked to Haunebu flights in Argentina, as we shall soon see.

The Volga Germans are ethnic Germans who colonized and historically lived along the Volga River in the region of southeastern European Russia around Saratov and to the south. Recruited as immigrants to Russia in the 18th century, they were allowed to maintain their German culture, language, traditions and churches. They were a range of Christian denominations: Lutheran, Reformed, Catholic, Moravian and Mennonite.

They even had their own independent republic within the

USSR called Volga German ASSR. It had a southeastern border with Kazakhstan. During World War II, after the German invasion of the Soviet Union in 1941, the Soviet government considered the Volga Germans potential collaborators, and deported many of them eastward, where thousands died. The Volga German Autonomous Soviet Socialist Republic was dissolved. It existed from 1924, shortly after the USSR was formed, to 1941.

The Deportation of the Volga Germans was the Soviet forced transfer of the whole of the Volga German population to Siberia and Kazakhstan on September 3, 1941. With secret orders, the trains were to stop at a station just outside of the region and the adult men were to be separated from their families. They were sent to Siberia or other areas to work in labor camps during the war. Of all the ethnic German communities in the Soviet Union, the Volga Germans represented the single largest group expelled from their historical homeland. After the war, the Soviet Union expelled a moderate number of ethnic Germans to the West. In the late 1980s and 1990s, many of the remaining ethnic Germans moved from

The short lived Volga German state within the USSR.

A map of German towns in the Volga River region of the USSR.

the Soviet Union to Germany.

However, a number of Volga Germans had emigrated to Argentina before WWII. In the early 1930s, following several years of drought in Europe, Volga German families emigrated to Argentina, Chile and Brazil. Many Volga German families who had settled earlier in the rural areas of La Pampa Province resettled to towns and villages like Jacinto Aráuz in search of work. This is the area of Argentina just south of Buenos Aires, with the coastal port of Bahia Blanca as the most important town. The town of Jacinto Aráuz is inland, to the northwest of Bahia Blanca.

The Volga German City in Argentina

It is rumored that one of the towns in Argentina where a great deal of Haunebu activity once occurred Jacinto Aráuz.

A curious incident involving UFOs and Volga Germans

occurred in Argentina on May 12, 1962 when three truck drivers, Valentino Tomassini, Gauro Tomassini and Humberto Zenobi, were travelling on Argentina Route 35 from Bahia Blanca to the town of Jacinto Aráuz.

At 4:10 in the morning they saw an object on the ground in a field next to the road at a distance of about 100 meters from them. The object looked like a railroad car and was illuminated. As their truck came close to it, the object rose up and crossed the road at a height of about four meters. The craft's lights then went out and from the lower part of the craft came a reddish flame. As the craft began to take off the object suddenly divided into two parts, and the two parts flew off in different directions. The men were stunned and went immediately to the police department and local newspapers.

That this strange encounter happened just near the town of

A map showing important towns in Argentina including those with Volga Germans.

Jacinto Aráuz is quite curious, because this is no ordinary town in Argentina, this is a town that was founded and populated by Volga Germans in the 1930s where many people speak German, rather than Spanish. This would be a good town for Nazi war criminals to inhabit and a good town to recruit new German-speaking personnel to continue the secret struggle of former officers of the Third Reich. We might imagine that certain young men from Jacinto Aráuz may have ended up at the Antarctic bases in Neuschwabenland and the South Shetland Islands.

It is also interesting to note that Patagonia does have an indigenous native population that may have also been recruited on occasion by the breakaway civilization that was becoming the Third Power. If so, these Native American young men with dark hair and dark complexions may have been part of the crew on some of the Haunebu and other craft. Perhaps this is the reason that some UFO reports, and Men in Black reports, talk about brown-skinned oriental-looking men being the occupants of such craft.

With the Volga Germans in South America and their harsh treatment in the Soviet Union, it is easy to see how the surviving elements of the Third Reich deemed Russia to be their major enemy and saw the United States as an ally—as long as it was opposed to the Soviet Union. The postwar German influence in many South American countries is well known and much of their focus was on supporting fascist military dictatorships and anti-communist movements. We also see here the beginning of a détente between the surviving Germans and the USA. With Rheinhard Gellen's Project Paperclip scientists now in top positions at NASA and Redstone Arsenal in Alabama, the German integration into the American space program could cautiously move forward.

The Volga Germans also to fit into the larger Nazi plan of German expansionism known as Lebensraum. These early German settlers had pioneered the German expansionism that the "Living Space" had promoted. In fact, the Volga Germans were at the southern end of the mighty Volga River and it was the Nazis plan to control all the area on the way to the Volga, including the territory of the Volga Germans and the territory south to Armenia and Baku.

A Haunebu Over Baku

In his book *The CIA UFO Papers,* author Dan Wright reports on a number of curious UFO incidents from CIA documents that were kept confidential for decades and have only been released because of the important Freedom of Information Act. Other countries do not have such government-sanctioned legal avenues for obtaining data (most of it comes from local newspapers).

Wright says that a CIA document from 1955 refers to an incident in Baku, the capital and largest city of Azerbaijan. Formerly a city with a large population of Armenians and Volga Germans, Baku is also in the vicinity of the Volga German state, dissolved by the Soviets in 1941 when Germany invaded the east. At the time, many Armenians and Volga Germans lived in Baku and ran the oilfields there. The Germans believed that it was crucial to capture Baku and its oilfields in order to fuel their occupation of Russia and the Ukraine. Baku and Armenia were the southernmost border of the Nazi state of Reichskommissariat Kaukasus, which was Southern Russia and the Caucasus region, including Armenia and Azerbaijan. The borders of Turkey and Iran created the southern frontier of this never-realized German colony.

The Germans were unsuccessful in taking Baku and its oilfields and had to retreat. Hitler was furious. This was the end for the Third Reich and the Greater Germany had not expanded to the important southern frontier of the Caucasus mini-state that the Germans were to create. Yet, after the war there were many Nazi collaborators in Ukraine, Armenia and the Baku area and some of them were Volga Germans. Is this the reason for Haunebu activity in the region in 1953?

The CIA document begins by saying that in 1953, a disc-shaped UFO was seen at the Soviet city of Shakhty, a coal-mining town at a spur of the Donetsk Mountains in southwestern Russia. The document then says that on August 9, 1955 outside of Baku, Azerbaijan witnesses saw a "flying saucer" take off two miles or so in the distance. They described a glow around the craft as it rose up into the sky.[40]

So, here we may be looking at some Haunebu craft picking up special people, possibly Volga Germans, on the outskirts of the city of Baku and whisking them away to someplace else.

104

The Haunebu base in western Tibet is a possibility, and that base was probably still active in 1955. One might think that it was abandoned in the years after that as the Communists expanded their control over Tibet, a sparsely populated area with few roads and towns. But who knows? The Nazis would have supported the original government of Tibet, one of an enigmatic hierarchy of lamas of various sects, typically opposed to each other. At the time of WWII and immediately afterwards, there was effectively no Dalai Lama as the previous Dalai Lama had died in 1933 and the current Dalai Lama was only a small child.

That a network existed of former Nazis, including Volga Germans, Ukrainians, Armenians, Tibetans and others who were in contact for many years after the war officially ended seems to be the case. Some of these people were met from time to time by SS Black Sun officers who brought them to safety by landing in West Germany, Spain or Argentina with a literal ride in a Haunebu. The Haunebu trip to Baku might also have exchanged gold, cash, or something else of value, including passports and other documents. Clearly the Haunebu activity in the Baku area in 1955 was not to attack the oil facilities there but to possibly drop people and documents off at the site and to pick people up to take them elsewhere. A Haunebu II could pick up a dozen or more passengers and safely transport them to a staging area—probably one of the Secret Cities of the Black Sun.

Such a network involved more than just secret bases and secret technology. It needed real contact with the outside world and the international economy. This was accomplished by using Argentina, Paraguay and Chile as de facto Black Sun client states. The heavy German investment in these countries helped their cause as did the many Germans in these countries. The fascist leanings of these countries during WWII and after also helped the SS with their "secret occupations." But, what was also needed were ambassadors and high-profile negotiators who lived a normal life in the normal world. These people would be industrialists, writers, and ex-Nazis who had escaped from justice after WWII such as Otto Skorzeny. Miguel Serrano is another of these characters who will be discussed in a later chapter. First let us look at Operation Bolivar and early intelligence for the Germans during WWII in

South America.

Operation Bolivar: Nazis in South America

Operation Bolívar was the codename for the German espionage in Latin America during World War II. Set up by the Abwehr, Germany's military intelligence agency, Operation Bolivar was headquartered in Buenos Aires. Some of its operatives in Argentina were probably Volga Germans. According to Wikipedia, Operation Bolívar was under the operational control of Department VID 4 of Germany's Security Service, and was primarily concerned with the collection and transmission of clandestine information to Europe from Latin America. The Germans were successful in establishing a secret radio communications network from their control station in Argentina, and also used a courier system involving Spanish merchant vessels for the shipment of paper-form intelligence. German agents of Operation Bolivar also travelled on Spanish ships to and from Europe. When they landed at a port in Spain they would fly on to Germany.

The main intelligence agent in Operation Bolivar was a man named Johannes Siegfried Becker, whose codename was "Sargo." Becker was first sent to Buenos Aires in May 1940, with orders to commit sabotage, along with his partner, Heinz Lange (Jansen), who arrived in the country that same year. Originally, these agents were to be involved in intelligence and sabotage, but the Argentine authorities made the Germans stick to just intelligence and there was to be no sabotage on Argentine territory. Says Wikipedia:

> However, after protests from the German embassy in Argentina in August 1940 the objective of the operation was revised to one of espionage only. Becker and Lange were soon discovered by Argentine authorities, so they moved their operations to Brazil, where they met with Gustav Albrecht Engels (Alfredo), who was another German spy, and the owner of the General Electric Company in Krefeld. Engels was originally recruited by the Abwehr, the German military's intelligence agency, in 1939 to collect and transmit economy-related intelligence from the Western Hemisphere back to Germany. Engels

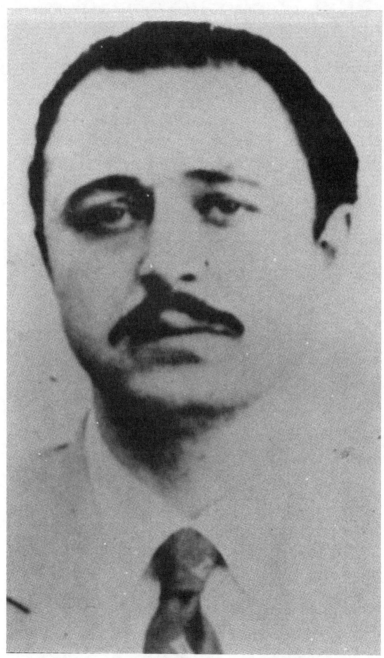

German spy Johannes Becker operated in South America.

thus established a radio station in São Paulo, the CEL, and used a radio transmitter owned by his electric company to relay information acquired by agents in both Brazil and the United States. When Becker arrived in São Paulo, he transformed Engels' operation into an organization that reported on all subjects of interest to German intelligence. This meant that, in addition to collecting economy-related information, the agents would collect information about shipping, war production, military movements in the United States, and political and military affairs in Brazil.

Although Bolívar was a Security Service project in origin, many of the agents responsible for collecting information were

A map of the countries in Latin America and Operation Bolivar.

part of the Abwehr. These agents used the many connections with Germans living in Argentina, Brazil, and other South American countries. German investment in and immigration to South America was considerable and it would only ramp up during and after the war. As an example, the beer brewery in Cuzco, Peru was built by the Germans during WWII and it is the only major industry in that city, even today. German investment was encouraged by all of the governments of South America, many of which actually supported the Nazis, though they wisely remained neutral during the war. Brazil eventually declared war against Germany in the last days of WWII, but it was largely a symbolic effort to please the Americans who were pressuring them for this action.

The Decline of the Abwehr and the Rise of the SS

Operation Bolivar in South America was an operation of the German military intelligence service known as the *Abwehr*. The *Abwehr* operated from 1920 to 1945. Under General Kurt von Schleicher the individual military services' intelligence units were combined and were centralized under his Ministry of Defense in 1929, forming the foundation for the more commonly understood manifestation of the *Abwehr*. Each *Abwehr* station throughout Germany was based on army district and more offices were opened in amenable neutral countries and in the occupied territories as the greater Reich expanded. The Ministry of Defense was renamed the Ministry of War in 1935. Vice-Admiral Wilhelm Canaris became the director with headquarters located in Berlin. Says Wikipedia:

> Before he took over the Abwehr on January 1, 1935, the soon-to-be Admiral Canaris was warned by Patzig of attempts by Himmler and Reinhard Heydrich to take over all German intelligence organizations. Heydrich, who headed the Sicherheitsdienst (SD) from 1931, had a negative attitude towards the Abwehr—shaped in part by his belief that Germany's defeat in the First World War was primarily attributable to failures of military intelligence, and by his ambitions to control all political intelligence-gathering for Germany.

Canaris was said to be a master of backroom dealings and he thought he knew how to deal with Himmler and Heydrich. Though Canaris tried to maintain a cordial relationship with Himmler and Heydrich, the antagonism between the SS and the Abwehr continued.

Wilhelm Canaris, director of the Abwehr.

Matters came to a head in 1937 when Hitler decided to help Joseph Stalin in the latter's purge of the Soviet military. Hitler ordered that the German Army staff should be kept in the dark about Stalin's intentions because they might warn their Soviet counterparts due to their long standing relations.

At this point special SS teams, accompanied by burglary experts from the criminal police, broke into the secret files of the General Staff and the Abwehr and removed documents related to German-Soviet collaboration. In order to conceal the thefts, the burglars started fires where they had broken in, including at the Abwehr headquarters in Berlin. Reinhard Heydrich was an SS Obergruppenführer and General der Polizei and one of the most powerful men in the Third Reich. He was assassinated in Prague in 1942 by Czech resistance fighters.

The Frau Solf Tea Party and the End of the Abwehr

On 10 September 1943, an incident occurred which came to be known as the "Frau Solf Tea Party." This tea party resulted in the dissolution of the Abwehr. Hanna Solf was the widow of Dr.

Wilhelm Solf, a former Colonial Minister under Kaiser Wilhelm II and the ex-Ambassador to Japan. Frau Solf was known to be involved in the anti-Nazi intellectual movement in Berlin. Members of her group were known as members of the "Solf Circle." At a tea party hosted by her on September 10, a new member was brought into the circle, a young Swiss doctor named Paul Reckzeh. Dr. Reckzeh was an agent of the Gestapo (Secret State Police), and he reported on the meeting, providing the Gestapo with several incriminating documents. The members of the Solf Circle were all rounded up on January 12, 1944. Eventually everyone who was involved in the Solf Circle was executed, except for Frau Solf and her daughter.

One of the persons executed was Otto Kiep, an official in the Foreign Office. Kiep had friends in the Abwehr, among whom were Erich Vermehren and his wife, the former Countess Elizabeth von Plettenberg, who were stationed as Abwehr agents in Istanbul. Both of them were summoned to Berlin by the Gestapo in connection with the Kiep case but in fear of their lives, they contacted the British Embassy and defected. Hitler had long suspected that the Abwehr had been infiltrated by anti-Nazi defectors and Allied agents, and the defection of Vermehren now confirmed it. Hitler and the SS also mistakenly believed the Erich and his wife absconded with the secret codes used by the Abwehr and gave them over to the British.

Hitler summoned Canaris for a final interview and accused him of allowing the Abwehr to "fall to bits." Canaris quietly agreed that it was "not surprising," as Germany was losing the war. Hitler fired Canaris on the spot and on February 18, 1944, Hitler signed a decree that abolished the Abwehr. Its functions were taken over by the Reichssicherheitshauptamt or RSHA (Reich Main Security Office) and overseen by SS-Brigadeführer and Generalmajor (Brigadier General) of Police Walter Schellenberg. With this action the SS took over all military intelligence and this strengthened Himmler's control over the military; the RSHA was also part of the SS. Schellenberg was captured by the British at the end of the war, was sentenced to six years in prison at the Nuremberg trials, but got out early because of a liver condition and died in Switzerland in 1952.

Canaris was given the empty title of Chief of the Office of Commercial and Economic Warfare. He was arrested on July 23, 1944, in the aftermath of the "July 20 Plot" against Hitler and executed in the spring of 1945, shortly before the end of the war.

English historian Hugh Trevor-Roper in his book *The Last Days of Hitler* (University of Chicago Press, 1992) says the Abwehr was, "rotten with corruption, notoriously inefficient, [and] politically suspect." He adds that it was under the "negligent rule" of Admiral Canaris, who was "more interested in anti-Nazi intrigue than in his official duties." According to Trevor-Roper, for the first two years of the war it was a "happy parasite" that was "borne along… on the success of the German Army."

As for German intelligence operations in South America—and elsewhere—that had been under the Abwehr, they were now under the SS. As of 1944, the SS controlled all military intelligence and this would include the missions that the Haunebu and other craft were to fly.

Back in South America

The setup of Operation Bolivar run by the Abwehr began with a network of radio stations in Argentina, Brazil, Paraguay and Chile. Says Wikipedia:

> One of the Abwehr spies in the United States that frequently traveled to Brazil to speak with Engels was Dušan Popov (Ivan), who was one of the most successful British double agents during the war. Other important Bolívar spies included the German naval and air attaché in Chile, Ludwig von Bohlen (Bach); the naval attaché in Rio de Janeiro, Hermann Bohny (Uncle Ernest); the military attaché in Buenos Aires, General Niedefuhr; and the naval attaché in Buenos Aires, Captain Dietrich Niebuhr (Diego), who headed the espionage organization in Argentina. In mid-1941, Herbert von Heyer (Humberto) joined the organization to provide maritime intelligence.
>
> Significant German espionage activity in Brazil ended in March 1942, when Brazilian authorities rounded up all suspected enemy agents. Becker was not in country,

however, having returned to Germany to meet with his superiors. It was during this time that Becker was put in charge of all German espionage activities in South America, all of which would center around radio communications, and ordered to make Buenos Aires his control station for communicating directly with Berlin, while also opening up smaller stations in other South American countries, which would relay information to the control station only. Heinz Lange, who had escaped Brazil to Paraguay before the arrests, was ordered to organize a spy network in Chile, and Johnny Hartmuth (Guapo), a Department VID 2 agent who had also escaped Brazil, was sent to organize a network in Paraguay. Furthermore, an agent named Franczok (Luna), was put in charge of the radio network that was to be established.

In February 1943, after considerable difficulty, Becker managed to return to Argentina as a stowaway on a ship traveling from Spain to Buenos Aires. Lange, Hartmuth, and Franczok, who airmailed one transmitter to Paraguay before leaving Brazil, set up a temporary station at Asunción, and reestablished contact with Berlin. However, after receiving Becker's orders, Franczok moved to the new control station in Buenos Aires in May 1943, Lange proceeded to Chile, and Hartmuth was left in Paraguay. Becker hoped to establish clandestine radio stations in every South American republic, but he was successful only in Paraguay, Chile, and Argentina.

The Nazis had always had a great interest in South America and its resources. South America had large pockets of German immigrants most of whom supported Germany in all of its efforts, including the colonization of South America. South America has tremendous resources in the mining and farming industries but was lacking in industrial manufacturing. This would change after the war, with the help of German companies who invested heavily in Argentina and Chile. What remained of the Third Reich could resettle in this southern third of South America where there was plenty of land, forests, mountains and lakes.

As the war was winding down, Argentina, Chile and Paraguay were still neutral (Brazil declared war against Germany at the very end of the war) and even in pro-Germany Argentina, the authorities arrested most of the German agents operating in the country in mid-1944, ending a great deal of Operation Bolívar activity. When discussing Operation Bolivar and its radio transmissions, it is often said that the information collected during the operation is believed to have been more useful to the Allies, who intercepted many of the secret transmissions, than to Germany. Says Wikipedia:

> As result of information collected by American counter-intelligence agencies and given to the Chilean government by the State Department, a number of the more active agents of the Chilean ring were arrested in the fall of 1942. Enough escaped, however, to permit von Bohlen to rebuild another network, known as the PQZ group. When von Bohlen went back to Germany late in 1943, his group was sufficiently well organized so that he could leave it, as well as a large sum of money and equipment, in the hands of Bernardo Timmerman, who carried it on until his arrest in February 1944. When Timmerman was arrested, the espionage rings in Chile were "smashed," but again some Germans managed to escape to Argentina, where they continued operating.
>
> In addition to revealing the identities of German spies and sympathizers, the interception of clandestine traffic allowed the Allies to maintain continuity on the agents operating in the Western Hemisphere. This information led to a number of arrests, the most celebrated at the time being that of Osmar Alberto Hellmuth on November 4, 1943. An Argentine naval officer, Hellmuth, unbeknownst to Argentina, was a German collaborator. His control, Hans Harnisch (Boss), claimed to be the personal representative of Heinrich Himmler and had extensive contacts in the highest reaches of the Argentine government. As a result of negotiations between Harnisch and various Argentine officials, including President Pedro Pablo Ramírez and various cabinet ministers, Hellmuth was appointed

Argentine consul in Barcelona. This appointment served to cover his actual mission: to proceed to Germany to assure that country that Argentina had no intention of severing relations with her. He was also to confer with the Security Service and other German officials on matters of mutual interest and was to obtain German permission for his return to Argentina from Sweden on the Argentine tanker Buenos Aires, carrying a load of German-supplied weapons.

Most of the details of this planning were known to the Allies through intercepted Bolívar radio traffic. As a consequence, when the *SS Cabo de Hornos*, aboard which Hellmuth was traveling to Spain, made a routine stop at Trinidad, British authorities arrested him. Argentina made a formal protest to Britain. When the ramifications of the affair were learned, however, there was a change in position. The Argentine minister of foreign affairs instructed his ambassador in London, on December 17, 1943, to inform Great Britain that Hellmuth's appointment had been cancelled and that if the British would release Hellmuth, his letters patent would also be cancelled and the British could then do with him as they saw fit.

The final note on Operation Bolivar, an operation controlled by the SS since 1944, is that information gained by US Naval Intelligence during the war, including various operatives was used by the State Department in preparing a case against the Peronista government of Argentina regarding its wartime support of the Axis in early 1946. At this time it was largely thought that the German intelligence operations had come to a halt in South America. But we now know that this was not the case. The SS continued to have intelligence operations throughout South America after the war and parts of Argentina, Paraguay and Chile became what we could call de facto colonies of the SS and the remnants of the Third Reich.

The SS Takes Control of the Black Reich

As the Third Reich dissolved and territory continued to be lost the SS was now in control of all of the intelligence and security

SS officers in black uniform, Himmler standing in front.

apparatus of what was left of the Nazi war machine. This was an astonishing array of: submarines; submarine bases; long-range aircraft; secret airbases in far-flung parts of the world; Haunebu and other special craft; plus electromagnetic pulse-type weapons that could disable or confuse trucks and aircraft (but not submarines, I theorize).

The SS had immense power at the end of the war and Hitler had demanded that the Third Reich fight to the last man, precipitating a bloodbath on his own country. But while the SS commanded these many airbases and submarine bases—some of them secret, some not—plus the military hardware associated with them, they had virtually lost all of the territory in their homeland and had no army left to fight for them. All they could do was escape to the various safety zones that they had carefully invested in during the years leading up to the war.

And so we can see how the Haunebu and the Vril were

116

important in the last months of the war and the years afterward in moving people, passports and financial documents out of Germany to Spain, Spanish possessions and South America. Others traveled by submarine to the Canary Islands, Antarctica and ultimately on to South America. The SS could only take a small elite from their ranks with them as they vacated the Reich. The big enemy, the Soviet Union, had vanquished the Germans and the elite within the SS would carry on with their high tech toys.

But what did they really have? The idea of fighting the world powers from Antarctica and Argentina was not going to work and therefore the weapons that all of their machines had were largely unnecessary. The submarines used by the Black Fleet after the war were probably the most heavily armed, but even they knew that they needed only defensive armaments, which would include some torpedoes. It seems that all of the dark U-boats after the war removed the deck guns that were clearly signs of a warship. When U-boats surrendered in Argentina, they were all missing their deck guns—they were no longer necessary.

The long-range aircraft, though originally designed as bombers, would not be going on any bombing missions. Any such mission would have been a one-way suicide mission with the complete loss of a very valuable plane, one of their few assets. Therefore, they were probably used on only a few long-range flights at the end of the war. Destinations such as Tibet, northern Norway, Greenland, Spanish Sahara and Argentina have been suggested.

The Strange Case of Hans-Ulrich Rudel

Hans-Ulrich Rudel (July 2, 1916—December 18, 1982) was a German ground-attack pilot during World War II. Rudel was the most decorated German serviceman of World War II, being the sole recipient of the Knight's Cross with Golden Oak Leaves, Swords, and Diamonds in January of 1945. Rudel typically flew the Junkers Ju 87 "Stuka"

Hans-Ulrich Rudel in 1943.

A Ju 87 equipped with anti-tank cannons photographed in 1943.

dive bomber and his missions were exclusively on the Eastern Front.

Rudel is credited during the war with the destruction of 519 tanks, as well as one battleship, one cruiser, 70 landing craft and 150 artillery emplacements. He claimed 11 aerial victories (earning flying ace status) and the destruction of more than 800 vehicles of all types. He flew 2,530 ground-attack missions exclusively on the Eastern Front, usually flying the Junkers Ju 87 "Stuka" divebomber as noted aboe. The Junkers Ju 87 "Stuka" was typically fitted with a single anti-tank cannon directly beneath the cockpit of the fighter. Photos of armed Haunebu craft have this very same anti-tank cannon on them, mounted beneath the craft.

Rudel surrendered to US forces on May 8, 1945 and was relcased after a period of time. In 1948 he emigrated to Argentina and began to network with all of his former Nazi buddies. IIe was a committed and unrepentant National Socialist, and founded the "Kameradenwerk" (literally "comrades work" or "comrades act"), an organization that helped Nazi fugitives escape to Latin America and the Middle East. Together with another Nazi in South America, Willem Sassen, Rudel helped shelter Josef Mengele, the notorious former SS doctor who practiced at Auschwitz. Rudel worked as an arms dealer and a military advisor to the regimes of Juan Perón in Argentina, Augusto Pinochet in Chile, and Alfredo Stroessner in Paraguay. These three nations in particular were suspected by US intelligence to be collaborating with the Nazis—before and after the war—and because of Rudel's activities, he was placed under observation by the newly formed Central Intelligence Agency. It is from this early period of the CIA following Rudel around South America that the curious CIA UFO files originate from.[40]

Says Wikipedia about some of the activities of Rudel's in South America:

> Prominent members of the "Kameradenwerk" included SS officer Ludwig Lienhardt, whose extradition from Sweden had been demanded by the Soviet Union on war crime charges, Kurt Christmann, a member of the Gestapo sentenced to 10 years for war crimes committed at Krasnodar, Austrian war criminal Fridolin Guth, and the German spy in Chile, August Siebrecht. The group maintained close contact with other internationally wanted fascists, such as Ante Pavelić, Carlo Scorza and Konstantin von Neurath. In addition to these war criminals that fled to Argentina, the "Kameradenwerk" also assisted Nazi criminals imprisoned in Europe, including Rudolf Hess and Karl Dönitz, with food parcels from Argentina and sometimes by paying their legal fees.
>
> In Argentina, Rudel became acquainted with notorious Nazi concentration camp doctor and war criminal Josef Mengele. Rudel, together with Willem Sassen, a former Waffen-SS and war correspondent for the Wehrmacht, who initially worked as Rudel's driver, helped to relocate Mengele to Brazil by introducing him to Nazi supporter Wolfgang Gerhard. In 1957, Rudel and Mengele together travelled to Chile to meet with Walter Rauff, the inventor of the mobile gas chamber.

In Argentina, Rudel lived in Villa Carlos Paz, roughly 36 kilometers (22 miles) from the populous Cordoba City, where he rented a house and operated a brickworks. There, Rudel wrote his wartime memoirs *Trotzdem* ("Nevertheless" or "In Spite of Everything"). This book was later re-edited and published in the United States under the title, *Stuka Pilot*. The book supported the

German invasion of the Soviet Union, which Rudel said was really the enemy in WWII, not the West.

The book was published in November 1949 by Dürer-Verlag in Buenos Aires, essentially a Nazi publishing company located in Argentina. They published a number of Nazi books and employed Germans who had been Nazis during the war who were living Argentina. Rudel seems to have started the publishing company in 1947 and it went bankrupt in 1958.

It is interesting that Rudel lived outside of Cordoba because this is a special area in Argentina. Cordoba is an industrial city in the north-central part of the country—and it is famous for Haunebu UFO activity.

This area is famous for a series of UFO disk sightings in the 1960s and early 1970s that occurred near the small city of Capilla del Monte, a short drive to the north of Cordoba. This small city is the center of Haunebu activity in northern Argentina and is well known for this. It is even a tourist destination because of the frequent sightings of flying saucers at the mountain on the edge of the city called Cerro Uritorco. Says Wikipedia about the town and the UFO phenomena centered on the nearby mountain:

> Capilla del Monte is a small city in the northeastern part of the province of Córdoba, Argentina, located by the Sierras Chicas mountain chain, in the northern end of the Punilla Valley. It has about 11,281 inhabitants as per the 2010 census.
>
> The main tourist attraction in the area is the Cerro Uritorco, a small mountain only three kilometers from the city, famed around Argentina as a center of alleged paranormal phenomena and UFO sightings. The city also features the tall El Cajón Dam and its large reservoir. Capilla del Monte was founded on October 30, 1585. Its name means "Mount's Chapel" in Spanish.

Capilla del Monte is a hotbed of UFO disk activity, which apparently continues to this day. In 1975 an unnamed newspaper reporter had photos of the disk craft taken at Cerro Uritorco on his camera and he was now at the Peruvian town of Sicuani with his

120

A Haunebu craft photographed at Cerro Uritorco in Argentina.

undeveloped film investigating other UFO sightings. He met with a Peruvian newspaper reporter working for the Lima newspaper *Ultima Hora* named Anton Ponce de Leon.

The photographer told Ponce de Leon that two men in black suits had come to his hotel when he was not there and told the clerk that they were his friends and asked to wait in the room for the photographer's return. Instead they ransacked his room, apparently looking for the film, and then left. The photographer had actually given the film to a friend in town to develop.

Later he met the men in black suits at the hotel lobby where they demanded his film. He told them the film had already been picked up and they left, but he found that his room continued to be entered and papers that were on a table were taken. He eventually fled the town in a taxi to Cuzco and then flew to Lima where the photos were ultimately published in an issue of *Ultima Hora*.[31]

Rudel and the Third Power in South America

It seems that Rudel was an early SS operative paving the way for more Nazi activity in South America, including the manufacture of parts for the Haunebu and even the complete manufacture of a Haunebu or Vril craft from the original plans. German investment

An SS dager inscribed with the motto: "My Honor is Loyalty."

in Cordoba was big immediately after the war, with Volkswagen and Mercedes Benz investing heavily in automobile plants in the city. Therefore, Rudel seems to have been using his brick factory for a Haunebu factory in Argentina's second largest town (or near to it). Rudel was writing books, allegedly running a brick factory, advising Juan Peron and Alfredo Stroessner, both supposedly very close friends, and networking with Nazis all over the world. Says Wikipedia:

> With the help of Perón, Rudel secured lucrative contracts with the Brazilian military. He was also active as a military adviser and arms dealer for the Bolivian regime, Augusto Pinochet in Chile and Stroessner in Paraguay. He was in contact with Werner Naumann, formerly a State Secretary in Goebbels' Ministry of Public Enlightenment and Propaganda in Nazi Germany. Following the Revolución Libertadora in 1955, a military and civilian uprising that ended the second presidential term of Perón, Rudel was forced to leave Argentina and move to Paraguay. During the following years in South America, Rudel frequently acted as a foreign representative for several German companies, including Salzgitter AG, Dornier Flugzeugwerke, Focke-Wulf, Messerschmitt, Siemens and Lahmeyer International, a German consulting engineering firm. Rudel's input was used during the development of the A-10 Thunderbolt II, a United States Air Force aircraft designed solely for close air support, including attacking ground targets such as tanks and armored vehicles.
>
> According to the historian Peter Hammerschmidt, based on files of the German Federal Intelligence Service and the US Central Intelligence Agency (CIA), the BND, under the cover-up company "Merex," was in close contact

with former SS and Nazi Party members. In 1966, Merex, represented by Walter Drück, a former Generalmajor in the Wehrmacht and BND agent, helped by the contacts established by Rudel and Sassen, sold discarded equipment of the Bundeswehr (German Federal armed forces) to various dictators in Latin America. According to Hammerschmidt, Rudel assisted in establishing contact between Merex and Friedrich Schwend, a former member of the Reich Main Security Office and involved in Operation Bernhard. Schwend, according to Hammerschmidt, had close links with the military services of Peru and Bolivia. In the early sixties, Rudel, Schwend and Klaus Barbie, founded a company called "La Estrella," the star, which employed a number of former SS officers who had fled to Latin America. Rudel, through La Estrella, was also in contact with Otto Skorzeny, who had his own network of former SS and Wehrmacht officers.

Rudel returned to West Germany briefly in 1953 and was the top candidate for the far-right German Reich Party but was not elected to the Bundestag. He moved back to Germany again and in 1977 he became a spokesman for the neo-Nazi political party German People's Union. Rudel died in West Germany in 1982.

Hans-Ulrich Rudel was a little-known Nazi and SS agent after the war, but his name crops up again and again. Rudel is clearly one of the top ex-Nazis that was running a "breakaway civilization" in industrial cities in three different countries and supplying the military of these countries with used weapons from West Germany. Rudel worked closely with Skorzeny and Peron, plus with Stroessner and Pinochet. One would expect there to be Haunebu bases in Capilla del Monte near Cordoba, and in several areas of the Patagonian region, including somewhere near Bariloche. One would expect there to be a Haunebu saucer base in Paraguay, probably in the western part of the country where huge ranches and numerous private airfields exist. In Chile we might expect there to be a Haunebu base and possibly a submarine base in the southern fjords of this long, thin country. Such a base would probably have been built during the war, and not after it. The arid

123

north of Chile is the Atacama Desert and a possible base for a Haunebu, though probably such a base would be in the Argentina portion of the Atacama, possibly near the Bolivian border.

If the Haunebu is a real aircraft and the UFO activity in South America is real, then Rudel was part of it. He must have taken a number of trips on a Haunebu and may have even been one of the pilots; it was certainly within his training. Early Haunebu craft had the anti-tank cannon on them but this was clearly discontinued after the war. One might even expect that Peron, Stroessner and other heads of state took a ride on a Haunebu. Such a flight would have been to impress these leaders with the technology created by the Third Reich. The SS was now in possession of this technology. But what could it do for these South American countries? As the years went on, the Nazis in South America became part of the mainstream society in these countries. There was widespread acceptance by the populations that there were Nazis in their countries. Some countries, like Paraguay, never tried to deny that were still working with Nazis after the war.

Rudel moved back and forth from South America to Europe and the Middle East. He met with heads of state and in the 1950s befriended Savitri Devi, a writer and proponent of Hinduism and Nazism, and introduced her to a number of Nazi fugitives in Spain and the Middle East. We will discuss Savitri Devi in the next chapter.

Indeed, we see that there is whole host of questionable characters active in South America after the war including Rudel, Skorzeny, Kammler and others. In fact, these former military men were well-trained and capable SS officers who were loyal to the Third Reich and the SS to the very end. Let us look at some of the other curious characters in South America, one of whom, Miguel Serrano, was a very strange person indeed.

124

Chapter 4

The Strange World of Miguel Serrano

Twenty degrees of solitude
Twenty degrees in all
All the dancing kings and wives
Assembled in the hall
Lost is a long & lonely time
Fairy Sybil flying
All along the all along
The mountains of the Moon.
—*The Mountains of the Moon*, Grateful Dead

There are a number of strange characters in South America that may have had something to do with the secret factories in Argentina and Chile, and none are stranger than Miguel Serrano. He was highly regarded in Chile and may have known many of the secrets of the Vril and Haunebu craft.

Miguel Joaquín Diego del Carmen Serrano Fernández, whose shortened name is Miguel Serrano, was born in Chile to European parents on September 10, 1917. He died on February 28, 2009 at the age of 92. He was a Chilean diplomat, author, occultist, and fascist political activist. Like many Europeans in Argentina, Paraguay and Chile, Serrano became a Nazi sympathizer in the late 1930s, and after the war he became a prominent figure in the neo-Nazi movement and a famous exponent of "Esoteric Hitlerism." We will see shortly how Esoteric Hitlerism involves the Haunebu and secret Nazi bases in use after the war in Antarctica and presumably Argentina.

Serrano was the child of a wealthy Chilean family but he was orphaned when his mother, Berta Fernández Fernández, died when he was five years old, and his father, Diego Serrano Manterola, died three years later. He had two younger brothers and a sister, all of whom were then raised by his paternal grandmother, Fresia Manterola de Serrano, moving between a Santiago townhouse and a 17th-century country mansion in the Claro Valley.

He studied at the Internado Nacional Barros Arana in Santiago where he developed an interest in writing and far-right politics. He became an activist in the Chilean Nazi movement starting around 1938.

Says Wikipedia:

During the Second World War, in which Chile remained neutral, Serrano campaigned in support of Nazi Germany and promoted anti-Semitic conspiracy theories through his own fortnightly publication, *La Nueva Edad* [*The New Era*]. In 1942, he joined an occult order founded by a German migrant which combined pro-Nazi sentiment with ceremonial magic and kundalini yoga. It presented the Nazi German leader Adolf Hitler as a spiritual adept who had incarnated to Earth as a savior of the Aryan race and who would lead humanity out of a dark age known as the Kali Yuga. Serrano became convinced that Hitler had not died in 1945 but had secretly survived and was living in Antarctica. After visiting Antarctica, Serrano travelled to Germany and then Switzerland, where he met the novelist Hermann Hesse and psychoanalyst Carl Jung; in 1965, he published a reminiscence of his time with the pair.

In 1953, Serrano joined the Chilean diplomatic corps and was stationed in India until 1963, where he took a keen interest in Hinduism and wrote several books. He was later made ambassador to Yugoslavia and then Austria, and while in Europe made contacts with various former Nazis and other far-rightists living on the continent. Following Chile's election of a Marxist President, Salvador Allende, Serrano was dismissed from the diplomatic service in 1970. After Allende was ousted in a coup and Augusto Pinochet took power, Serrano returned to Chile in 1973. He became a prominent organizer in the Chilean neo-Nazi movement, holding annual celebrations of Hitler's birthday, organizing a neo-Nazi rally in Santiago, and producing a neo-Nazi political manifesto. He wrote a trilogy of books on Hitler in which he outlined his view of the Nazi leader as an avatar. He remained in contact with neo-Nazis elsewhere in the world and gave interviews to various foreign far-right publications.

Between 1929 and 1934, Serrano studied at the Internado Nacional Barros Arana—a school that had been heavily influenced by Prussian staff members who had arrived in the late 19th century, and Serrano attributes his later Germanophilia to this early exposure to German culture at school. Serrano moved in literary circles and a close friend of his was another student, 18-year old Hector Barreto, a poet and socialist. Barreto was killed in a street brawl with uniformed Nacistas, members of the National Socialist Movement of Chile, a fascist group inspired by the example of the Nazi Party in Germany. Rudolf Hess in Germany was attending similar street brawls at this same time in Germany, and Serrano would later visit him in Spandau Prison.

The tragic event of his friend's death encouraged Serrano's involvement in left-wing politics and he began to take an interest in Marxism and the Chilean Marxist movement. He wrote articles for Chilean leftist journals like *Sobre la marcha, La Hora,* and *Frente Popular*. His uncle, the poet Vicente Huidobro, encouraged him to join the left-wing Republicans in the ongoing Spanish Civil War, but Miguel did not want to leave his intellectual studies.

Instead, Serrano became critical of Marxism and was drawn to the Chilean fascist movement known as the Nacistas after a failed coup in September of 1938. By July 1939, Serrano was publicly associating himself with the Nacista movement, now organized as the Popular Socialist Vanguard. Serrano began writing for the group's journal, *Trabajo*, and accompanied its leader, Jorge González von Marées, on speaking tours around Chile. At the outbreak of WWII, Serrano expressed support for Nazi Germany; from July 1941 he launched a fortnightly pro-Nazi publication, *La Nueva Edad*. Through these publications, Serrano developed close links with the local Nazis and the personnel at the German

Embassy in Chile.

According to the historian Nicholas Goodrick-Clarke in his book *Black Sun*,[28] Serrano had initially shown little interest in Nazi attitudes towards the Jewish people, but he became increasingly interested in anti-Semitic conspiracy theories about Jews manipulating world events. Goodrick-Clarke says that two Chilean artists gave him a Spanish language translation of the *Protocols of the Elders of Zion*, a text originally published in Russia purporting to expose this alleged international Jewish conspiracy. According to Goodrick-Clarke, it was this discovery of the *Protocols* which "marked a crucial point in the development of Serrano's Nazism." Serrano began printing excerpts from the *Protocols* in *La Nueva Edad* starting in November of 1941.

Goodrick-Clarke says that Serrano also developed an interest in forms of religious or spiritual practice, including both Western esotericism and Hinduism. In late 1941, Hugo Gallo, who was the cultural attaché at the Italian Embassy, suggested that Serrano could support the German and Italian war effort not just through his publications, but also on the etheric Inner Planes. Gallo then introduced Serrano to an esoteric order, in a sense a secret society, that was mystical in nature and sympathetic to Nazism.

Serrano later claimed that this secret society had been founded in the early 1900s by a German migrant known as "F. K." Serrano says he was initiated into the group in February 1942. The mysterious F. K. claimed that the group owed its allegiance to a secretive Brahmin elite of Masters who resided in the Himalayas. Serrano said the order practices combined kundalini yoga with ceremonial magic and expressed a pro-Nazi position during WWII. It espoused a belief in an astral body that could be awakened through various rituals and meditative practices.

Serrano says that the group revered Adolf Hitler as the savior of the Aryan race and presented him as a *shudibudishvabhaba*, an initiate of immense willpower who had voluntarily incarnated onto Earth to assist in the overthrow of the current Kali Yuga, a present dark age for humanity. F. K. claimed that through the astral realm, he was able to establish a connection with Hitler, during which they had various conversations.

With the defeat of Germany in the spring of 1945, F. K. told

Serrano that Hitler was not dead and that he remained in astral contact with Hitler. Serrano and others in South America came to believe that Hitler had not died in the bunker in Berlin but had escaped Germany to Antarctica, where he was living in a secret base beneath the ice in Neuschwabenland.

The Latin press had already been suggesting as much and in the early years after the war there was much speculation as to whether Hitler was really dead or whether he had escaped to South America or elsewhere. In 1947, a book was published in Buenos Aires called *Hitler est vivo* by Ladislao Szabó.[19] Szabó's book alleged that a U-boat convoy had taken Hitler to safety to a secret base in Neuschwabenland, Germany's Antarctic territory. This book was a big influence on Serrano and from late 1947 into the spring of 1948, Serrano travelled in Antarctica as a journalist with the Chilean Army where they set up a base in western Antarctica. On his return to Santiago Serrano wrote his own short book, *La Antártica y otros Mitos,*[20] which was published at the end of 1948. In this 52-page booklet, Serrano repeated Szabó's claims about Hitler's escape to Antarctica.

In 1951, Serrano travelled to Europe, and while in Germany he visited various sites associated with the Nazi Party, including Hitler's Berlin bunker, Hitler's Berghof home, and Spandau Prison, where prominent Nazis such as Rudolf Hess and Karl Donitz were being imprisoned. After his German travels Serrano visited Switzerland, where he met the psychoanalyst Carl Jung and was befriended by the writer Hermann Hesse.

Back in Santiago, like other members of his family, Serrano joined the Chilean diplomatic corps in 1953. He hoped to be posted to India, which he considered to be the motherland and a source of great spiritual truths. He got his wish and was sent to India where he remained until 1962.

In this period, he visited many Hindu temples and searched for

The 1947 book *Hitler is Alive* published in Buenos Aires.

129

evidence of the secretive Brahmanical order into which F. K. had alleged initiation. In his role as a diplomat, he met various prominent figures, including Jawaharlal Nehru, Indira Gandhi, and the 14th Dalai Lama. While in India Serrano wrote and published two books: *The Visits of the Queen of Sheba* (1960), which had a preface by Jung, and *The Serpent of Paradise* (1963), which discussed his experiences in the country. Serrano had engaged in further correspondence with Jung between 1957 and 1961. In 1965 his book *C. J. Jung and Hermann Hesse: A Record of Two Friendships* was published.

He also met Savitri Devi Mukherji, another mystical unrepentant Nazi, famous for writing the book *The Lightning and the Sun*, a book that would be a profound influence on Serrano. More on Savitri Devi in a moment.

Serrano left India in 1962 and became the Chilean ambassador to Yugoslavia until 1964. He then served as Chile's ambassador to Austria until 1970, at which time he lived in Vienna. During the

Miguel Serrano in India, 1957.

Miguel Serrano with Jawaharlal Nehru, 1957.

latter posting, he also represented Chile at the International Atomic Energy Agency and the United Nations Industrial Development Organization, both of which were based in Vienna.

While in Europe, he sought out a number of individuals linked to Nazism and to the far right including visits to the Ahnenerbe co-founder Herman Wirth and the designer, writer and occultist Wilhelm Landig. Serrano established friendships with a number of individuals involved in the old Nazi movement, including Otto Skorzeny, Hans-Ulrich Rudel, Léon Degrelle, Marc "Saint-Loup" Augier, and Hanna Reitsch. He also discussed ideas with the French ancient astronaut proponent Robert Charroux whose books included a lot of material about South America.

In 1970 the Marxist Salvador Allende was elected president of Chile, and later that year Serrano was dropped from the country's diplomatic service. Rather than return to Chile, Serrano moved to Switzerland, renting an apartment in the Casa Camuzzi at Montagnola—the same house where Hermann Hesse had lived from 1912 to 1931.

The Lightning and the Sun

The loss of his diplomatic position, coupled with the establishment of a Marxist government in Chile, led Serrano to

131

Savitri Devi in 1937.

take a revived interest in Nazism. He began reading a number of recently published books that purported to identify links between Nazism and occultism. In 1973, his book *El/Ella: Book of Magic Love* was published.

Salvador Allende was ousted in a September 1973 coup and a right-wing military regime under Augusto Pinochet took power; therefore Serrano returned to Chile. However, the new Pinochet administration was not interested in his neo-Nazi and Esoteric Hitlerist ideas and would not hire him as a foreign diplomat.

During this time Serrano began producing a trio of books that came to be known as his "Hitler Trilogy" of Esoteric Hitlerism: *El Cordón Dorado: Hitlerismo Esotérico* (*The Golden Cord: Esoteric Hitler*, 1978), *Adolf Hitler, el Ultimo Avatãra* (*Adolf Hitler, the Last Avatar*, 1984), and *Manú: "Por el hombre que vendra"* (*Manu: For the Coming Man*, 1991). These books were heavily influenced by the 1958 book *The Lightning and the Sun* by Savitri Devi Mukherji. Serrano had sought her out in Calcutta when he lived there and was very impressed by her book which had just been published in that city.

Savitri Devi Mukherji was born Maximiani Julia Portas in France on September 30, 1905 (died October 22, 1982). In early 1928, she renounced her French citizenship and acquired Greek

132

nationality. While on a pilgrimage to Palestine during Lent in 1929, she decided that she was a Nazi.

In 1932, she travelled to India in search of a living pagan Aryan culture. Formally adhering to Hinduism, she took the name Savitri Devi ("Sun-rays Goddess" in Sanskrit). She volunteered at the Hindu Mission as an advocate against Judeo-Christianity, and wrote a booklet titled *A Warning to the Hindus* to offer her support for Hindu nationalism and independence. This was part of her effort to rally resistance to the spread of Christianity and Islam in India. She writes in her books that in the 1930s she distributed pro-Axis propaganda and engaged in intelligence gathering on the British in India.

In 1940, Devi married Asit Krishna Mukherji, a Bengali Brahmin with Nazi views who edited the pro-German Calcutta newspaper *New Mercury*. Devi and Mukherji gathered intelligence

Savitri Devi's husband AK Mukherji in Calcutta in 1939.

133

for the Axis cause using techniques such as entertaining Allied personnel, which gave Devi and Mukherji an opportunity to question them regarding military matters. The information gathered was passed on to Japanese intelligence officials.

After World War II, she travelled to Europe in late 1945 under the name Savitri Devi Mukherji as the wife of a British subject from India, under a British Indian passport. She stopped briefly in England, visited her mother in France, and then travelled on to Iceland where she witnessed the eruption of Mount Hekla. She then returned to England, before travelling to Sweden where she met with the famous Central Asian explorer Sven Hedin.

In June 1948 she took a train from Denmark to Germany, where she distributed many thousands of copies of handwritten leaflets encouraging the "men and women of Germany" to "hold fast to our glorious National Socialist faith, and resist!" Arrested for posting bills, she was tried in Düsseldorf on April 5, 1949 for the

134   Savitri Devi in 1947.

promotion of Nazi ideas on German territory subject to the Allied Control Council, and was sentenced to two years imprisonment. She only served eight months in Werl Prison, where she befriended fellow Nazi and SS prisoners, before being released and expelled from Germany. She went to France, staying in Lyon for several years.

In 1953 she went to Greece and obtained a Greek passport in her maiden name in order to re-enter Germany, and she began a pilgrimage, as she called it, of Nazi "holy" sites. She flew from Athens to Rome then travelled by rail over the Brenner Pass into "Greater Germany," which she regarded as "the spiritual home of all racially conscious modern Aryans." She travelled to a number of sites significant in the life of Adolf Hitler and the NSDAP (Nazi party), as well as German nationalist and heathen monuments, and recounted the voyage in her 1958 book *Pilgrimage*.

Also during this time, Savitri Devi became friends with the prominent Nazi Hans-Ulrich Rudel, discussed in the last chapter, and completed her manuscript of *The Lightning and the Sun* at his Spanish home in March of 1956. Because of her friendship with Rudel, she was able to meet a number of Nazi émigrés in Spain and the Middle East. She became marvelously funded by Rudel and other Nazis and made all sorts of trips to the Middle East and around Europe. In 1957 she stayed with Johann von Leers in Egypt and traveled across the Middle East on her way home to New Delhi, including stops in Beirut, Damascus, Baghdad, and Tehran. Returning to Europe in 1961 she stayed with Otto Skorzeny in Madrid.

In 1958 a Calcutta printer published her opus *The Lightning and the Sun,* in which she outlines her philosophy of history along with a critique of the modern world. It is a work synthesizing the Hindu philosophy of cyclical history with Nazi ideas. It contains biographies of Genghis Khan, Akhenaton and Adolf Hitler. The book is known for the author's claim that Adolf Hitler was an avatar of the Hindu God Vishnu.

Begun in 1948, completed in 1956, and first published in 1958 she said the book "could be described as a personal answer to the events of 1945 and of the following years." It is dedicated "To the god-like Individual of our times; the Man against Time;

135

the greatest European of all times; both Sun and Lightning: Adolf Hitler, as a tribute of unfailing love and loyalty, for ever and ever." It opens with quotations from *The Bhagavad Gita* and Rudolf Hess.

In *The Lightning and the Sun*, Savitri Devi attempts to weave Nazism with a cyclic view of history, arguing that time begins with a Golden Age and gradually decays through a Silver Age and Bronze Age into a final Kali Yuga, or Dark Age. She elucidates her concept of "Men in Time," "Men above Time," and "Men against Time" using the lives of Genghis Khan, Akhenaton, and Adolf Hitler respectively. Genghis Khan is used as an example of a "Man in Time" who exhibits Lightning (destructive) qualities and furthers historical decay. Akhenaton is used to illustrate a "Man above Time" who exhibits Sun qualities (creative/life-affirming) and seeks to transcend the process of historical decay. Adolf Hitler is used to illustrate a "Man against Time" who exhibits both "Lightning and Sun" qualities (destructive power harnessed for a life-affirming purpose) and seeks to fight historical decay by using violent, Dark Age methods to achieve a Golden Age state of existence.

In the final chapter, Savitri Devi expands further upon her cyclic view of history and argues that at the end of the Kali Yuga/ Dark Age, Kalki will appear and usher in a new Golden Age. She believed Hitler to have been sent as an avatar of the Hindu god Vishnu. She believed Hitler was a sacrifice for humanity that would lead to the end of the Kali Yuga that had been induced by the Jews.

After staying with Otto Skorzeny in Spain during 1961, she took employment teaching in France during the 1960s, spending her summer holidays with friends at Berchtesgaden in Germany. She retired from teaching in 1970 and Savitri Devi spent nine months at the Normandy home of her close friend Françoise Dior, working on her memoirs. She was welcomed at first, but her annoying personal habits of chewing garlic continually and not taking a bath during her entire stay, began to annoy Dior who encouraged Savitri Devi to return to India where her pension would go much farther. Savitri Devi flew from Paris to Bombay in June of 1971 and ultimately moved to an apartment in New Delhi,

136

Otto Skorzeny in the uniform of an SS officer. Did he control the postwar Reich?

where she lived alone, with a number of cats and a cobra.

By the late 1970s Savitri Devi had developed cataracts and her eyesight was rapidly deteriorating. A clerk from the French embassy in India looked after her, making regular house visits. She decided to leave India, returning to Germany to live in Bavaria in 1981 and then moved to France in 1982. She eventually died in 1982 in Sible Hedingham, Essex, England, at a friend's home.

The Lightning and the Sun had a huge impact on Miguel Serrano and was something of a bible to postwar Nazis of the 1960s. Essentially, Serrano was to add to Savitri Devi's work and continue along the theme of Hitler as an avatar of Vishnu coming to end the current Kali Yuga with a great deal of death and destruction. Serrano termed his philosophy Esoteric Hitlerism,

Savitri Devi with her cat Black Velvet in later years. Did she know about the Haunebu?

which he described as a new religious faith "able to change the materialistic man of today into a new idealistic hero," and as "much more than a religion: It is a way to transmute a hero into God." As I said, he believed that Hitler was an avatar of Vishnu, as did Savitri Devi.

In 1984 he published his second book in his Hitler Trilogy, the 643-page *Adolf Hitler, el* Último *Avatāra*[25] (*Adolf Hitler: The Last Avatar*), which is dedicated "To the glory of the Führer, Adolf Hitler." In this arcane work, Serrano unfolds his ultimate philosophical testament through elaborate esoteric and mythological symbolism. He insists that there has been a vast historical conspiracy to conceal the origins of evolved humankind.

The book opens with extragalactic beings who founded the First Hyperborea, a terrestrial but non-physical realm, which was neither geographically limited nor bound by the circles of reincarnation. The Hyperboreans were asexual and reproduced through "plasmic emanations" from their ethereal bodies. They could command vril power, the light of the Black Sun coursed through their veins, and they saw with the third eye. Serrano contends that the last documents relating to them were destroyed along with the Alexandrian Library, and that these beings have been misunderstood as extraterrestrials arriving in spaceships or UFOs. However, the First Hyperborea was immaterial and altogether outside our mechanistic universe. Here we see the influence, and somewhat of a rejection, of the ancient astronaut theories of Robert Charroux.

The First Hyperborea is under the jurisdiction of the Demiurge, an inferior godlet whose realm is the physical planet earth. The Demiurge had created a bestial imitation of humanity in the form of proto-human "robots"—the Neanderthals—and intentionally consigned these creatures to an endless cycle of involuntary reincarnation on the earthly plane to no higher purpose. The Hyperboreans recoiled in horror from this entrapment within the Demiurge's cycles. They themselves would take the *devayana*— the Way of the Gods—at death and return to the earth (as Bodhisattvas), but only if they are willing.

Determined to wage a heroic war to reclaim the Demiurge's deteriorating world, the Hyperboreans clothed themselves in

139

material bodies and descended to the Second Hyperborea, a ring-shaped continent around the North Pole.

During this Golden Age—or Satya Yuga—they magnanimously instructed the Demiurge's creations (the Black, Yellow and Red races native to the planet) and began to raise them above their animal condition. Then disaster struck; some of the Hyperboreans rebelled and intermingled their blood with the creatures of the Demiurge, and through this transgression Paradise was lost. Serrano refers to *Genesis 6.4*: "the sons of God came in to the daughters of men, and they bore children to them." By diluting the divine blood, the primordial miscegenation accelerated the process of material decay.

This was reflected in outward catastrophes and the North and South Poles reversed positions as a result of the fall of a comet or moon. The polar continent disappeared beneath the deluge and Hyperborea became invisible again. The Hyperboreans themselves survived, some taking refuge at the South Pole. Serrano regards the mysterious appearance of the fine and artistic Cro-Magnon Man in Europe as evidence of Hyperboreans driven southward by the Ice Age. In the then-fertile Gobi Desert, another group of exiled Hyperboreans established a fantastic civilization. Much of this material can be found in the books of Robert Charroux, whom Serrano met with in Paris and had many discussions on Atlantis, South America, the Gobi Desert and ancient astronauts. Charroux was unique among the 60s and 70s ancient astronaut proponents in that he constantly featured the Gobi Desert as an inland sea with an island in the middle of this Gobi Sea called the "White Island." It is interesting to note along these lines that Lake Baikal on the northern border of Mongolia and Russia is a deep, freshwater lake that has a population of seals—but is thousands of miles from any ocean. Charroux and Serrano believed this to be evidence of the inland Gobi Sea and the mysterious White Island, presumably populated by Aryan-Hyperboreans.

The world thus becomes the combat zone between the dwindling Hyperboreans and the Demiurge and his forces of entropy. But Serrano claims that the Golden Age can be reattained if the descendants of the Hyperboreans—the Aryans,—consciously repurify their blood to restore the divine blood-memory. Says

140

Serrano in *Adolf Hitler, el* Último *Avatãra*:

> There is nothing more mysterious than blood.
> Paracelsus considered it a condensation of light. I believe
> that the Aryan, Hyperborean blood is that—but not the
> light of the Golden Sun, not of a galactic sun, but of the
> light of the Black Sun, of the Green Ray.

A Journey Across the Ice to Meet the Führer

Serrano increasingly associated with old Nazis living in Chile
as well as with their neo-Nazi sympathizers. His trips to Europe
essentially stopped. In May of 1984 he attended the funeral of
Walter Rauff—a member of the Waffen SS who had played a role
in organizing the early stages of the Holocaust and who had fled
to Chile after the Second World War—and there gave the Nazi
salute. This important postwar Nazi funeral was attended by all
sorts of people including Mossad and CIA agents.

In 1986 Serrano published a political manifesto for Nazism in
the Southern Cone of South America. He began organizing annual
celebrations of Hitler's birthday at a rural retreat in Chile. At this
time it was still not generally known that Hitler had been living in
Argentina for decades on an island in a lake in Patagonia, although
rumors to this effect had been in the press for years.

In September 1993, he led a neo-Nazi rally in Santiago—
dressed in what had become his trademark black leather coat—in
honor of the Nazi Rudolf Hess and the Nacistas—early Chilean
Nazis killed by the government following their 1938 coup attempt.
Serrano continued to write and have books published in Santiago.
He died on February 28, 2009, after suffering a stroke in his
apartment in the Santa Lucía Hill sector of Santiago.

During his later years he wrote several books including an
autobiographical trilogy starting with the 1996 book *Memorias de
Él y Yo vol. I* (*Memories of Him and Me. Volume 1.*

This was followed by the 1997 book *Memorias de Él y Yo, vol.
II: Adolf Hitler e la gran Guerra.*[27] Serrano then published his final
book in the trilogy in 1998, *Memorias de Él y Yo vol. III, Misión en
los Transhimalaya* (*Memories of Him and Me. Volume 3. Mission
in the Transhimalaya*). This book is about his mission in India

and his search in the Himalayas for the original headquarters of the esoteric order to which he belonged in Santiago, and which supported Hitler and his revolution, starting Serrano on his life's path. He writes about his friendship with Jawaharlal Nehru, Indira Gandhi and the Dalai Lama. He also writes about his friendship with Carl Gustav Jung and other international personalities of those years.

A prolific writer, even in his older days, Serrano published the 72-page booklet *El hijo del viudo* (*The Son of the Widower*) in 2003. In this short work, Serrano makes a synthesis of all his work through a detailed and profound analysis of esoteric Christianity, the esotericism of Islam and that of the Hitlerite SS. Here we see how Serrano is indeed steeped in lore of the SS and the Black Sun and appears to be an initiate of this order.

Serrano continued to write, and his last one was the 2005, 44-page booklet *Maya: La Realidad Es Una Ilusión* (*Maya, Reality is an Illusion*), in which Serrano maintained that the Germans had mastered the art of the "double" and that the Rudolf Hess in Spandau Prison had been a double and was not the real Rudolf Hess.

Serrano died on 28 February 2009, after suffering a stroke in his

Miguel Serrano's hero, Adolf Hitler.

apartment in the Santa Lucía Hill sector of the capital, Santiago. According to Goodrick-Clarke, Serrano's "mystical Nazism" was "a major example of the Thulean mythology's successful migration to South America in the postwar period." He thought it "likely that old Nazis welcome[d] Serrano's enthusiasm and unswerving loyalty to their hero, Adolf Hitler," even if they found the Esoteric Hitlerist mythology that he promoted to be farfetched. On the other hand, Goodrick-Clarke thought that for younger neo-Nazis: "a coloring of pop mythology, Hinduism, and extraterrestrial Aryan gods adds sensational appeal to the powerful myths of elitism, planetary destiny and the cosmic conspiracy of the Jews."[28]

In a 2017 article in the Australian magazine *New Dawn* the British scholar Joscelyn Godwin discusses Serrano and his Theosophical and other beliefs. Says Godwin of Serrano's claim of an astral flight to meet Hitler during the 1947-1948 Chilean Antarctic expedition:

> In India he had friendly relations with Nehru, Indira Gandhi, and the Dalai Lama, whom Serrano was the first diplomat to welcome after his flight from Tibet. He kept company with sadhus and practiced yoga, complementing an already extensive knowledge of Western esoteric traditions. In Europe he enjoyed intimate visits with Hermann Hesse and Carl Gustav Jung, as well as with surviving members of Hitler's circle. Gradually he revealed his identity and mission as an "esoteric Hitlerist," elaborating a personal mythology and symbology around the idea that Hitler was the Tenth and Last Avatar of Hindu doctrine. In time, this became more and more mystical and apocalyptic, encompassing self-regeneration, Theosophical and Ariosophical prehistory, UFOlogy, and conspiracy theory.
>
> In an interview with *La Nación*, 3 April 2005, the 88-year-old Serrano replied to a journalist's question: "Antarctica is surely the frozen Atlantis. Imagine that under the Antarctic subterranean rivers have been discovered as long as the distance from Santiago to Talca (300 kms). Beaches under the Russian and Chinese bases. And the

warm streams. Hitler and the Germans would be in that Hyperborean zone, where there are subterranean walkways millions of years old, that go nobody knows where."

Serrano himself had visited the continent in summer 1947/48, as the only civilian member of an expedition to establish a Chilean military base. He wrote about his experiences several times, each time revealing a little more of their secret content. Finally, in the second volume of his autobiography (1997) he wrote this astonishing account of his meeting with Hitler, in a blend of UFO abduction imagery, astral travel, and a lecture on occult prehistory.

He then quotes from Serrano's *Memorias de Él y Yo, vol. II: Adolf Hitler e la gran Guerra*[27]:

It was the very night before our expedition departed when the following occurred. I found myself alone in the tent. A light illuminated the entrance, and there was someone there, dressed in black, tall, willowy. He gestured to me to get up. I got out of my sleeping bag with no difficulty and rose. We quickly sledded over the plain, now in complete darkness without a star in the sky. We could just see the lights of the frigate anchored in the bay. Suddenly it disappeared and we found ourselves before the mouth of a cavern in the glacier. The wall of ice opened in the depths, and we were at the center of a marvelous world, beside a lake surrounded by forests of trees of unknown species. Soldiers of the SS were occupied in maneuvers, or guarding a circular vehicle that might have been made of brilliant metal, though possibly it was a pure circular vibration. Some officials opened the way and led us up to the Disk. There were no doors or windows in the metal, or the metallic vibration, as if it was made of air. Once inside, the vehicle vibrated ever more intensely, so that I thought that they were my own vibrations, in tune with it. Even when inside, properly seated and firmly held by luminous belts, I could see everything that happened outside. We rose and rapidly plunged into the lake waters.

Right: This color photo of a Haunebu is said to have been taken in Berlin in 1935. From the Ettl documents.

A Haunebu in flight, unknown date. From the Ettl documents.

Left: A Haunebu on the ground. *Right:* A Haunebu in flight with a blimp, unknown date. From the Ettl documents.

Left: An early Haunebu in flight, unknown date. From the Ettl documents.

A Haunebu in flight, unknown date. From the Ettl documents.

Left: Nazi flying ace Hans-Ulrich Rudel.

Right: A Haunebu in flight, unknown date. From the Ettl documents.

Left: An interesting photo of a Haunebu with officers in the foreground. From the Ettl documents.

A Haunebu on the ground, unknown date. From the Ettl documents.

An amazing Haunebu III photo showing weapons and antennas plus soldiers in the field. One might think that this is a fake photo were it not for the overwhelming evidence for the existence of the Haunebu, Vril and Andromeda craft. From the Ettl documents.

A 727 pilot took this photo of a Haunebu in 1976 during a flight over the Brazilian Amazon,.

A photograph of what appears to be a Haunebu at Floridad, Uruguay taken on July 11, 1977.

An artist's painting of a Haunebu II inside of a hangar.

The floor of the main room at Wewelsburg Castle with the Black Sun insignia there since 1941. Himmler bought the castle in 1934 to be an academy for SS officers.

Above left: A rare color photograph of a Haunebu on the ground. *Above right:* A rare color photograph of a Vril craft in flight. *Below left:* A rare color photograph of a Vril craft seen from below.

A photograph of a flying disk very similar to the Haunebu taken near Kanab, Utah on June 10, 1964.

Left: An early Vril craft in flight with a car in the foreground. From the Ettl documents.

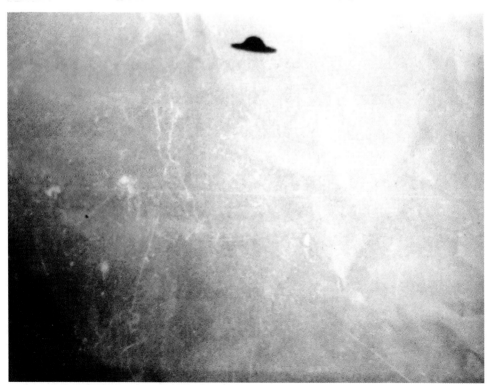

A photo of a Vril craft taken by Yamandu Lopez at Playa Shangrila in Uruguay on September 23, 1968 around 6:30 PM. He had brought his camera to the beach to take pictures of his children.

For a moment there was total darkness, then a universe reappeared, with a central sun and cities superimposed. This was the Hollow Earth, the Earth's Double. It was an Astral Earth, absolutely mental, *depending for its existence on my thought and on my continuing to think*. There the Fourth Reich was now installed and triumphant. I was led by young, martial officials who smiled affably at me into the presence of the *Führer*, the *Last Avatar*. I did not see his face, but I felt his reality, touching the depths of my being, my ever-present "I." And this is what he said to me:

> *"Heil, Sieg Heil! Benvenido!* You can become a part of my Last Battalion, the *Einherjar*, when you are purely astral, when you have sloughed off your physical body and regained your subtle Hyperborean form, learning your Master's practice to materialize and dematerialize your astral body at will. This is what the God-Men, the Supermen in Hyperborea, the Giants were dong when they were enamored of the daughters of men, the genetic robots, the plagiarisms of the Demiurge, and lost their Power and ability to develop a pure mental body…"[27]

So here we have a story from Serrano's early expedition to Antarctica where he has a mystical meeting with his hero, Adolf Hitler. It is part journey on a sled and part astral journey to a secret Nazi base in Antarctica where SS troops guard a flying saucer and Serrano is able to meet his hero, Adolf Hitler, if only briefly. This supposed astral journey to meet the Führer was in 1947/48 when Serrano was still a young man, only 30 years old. Did he have a meeting with Hitler at a later date and was he ever given a ride in a Haunebu?

It would seem that Miguel Serrano carried his beliefs in Hitler and the Third Reich right to his grave in 2009 and one wonders what secrets he might have known about the Haunebu, Antarctica and Hitler's survival that the mainstream did not know. If portions of the SS escaped in U-boats, long-range aircraft and Haunebu

Juan Perón as an Argentinian general in 1940.

flying saucers, then maybe Hitler was alive, living in Antarctica and later in Argentina, just as Serrano wrote. Except, Serrano never wrote about Hitler living in Argentina, and that may be because he knew that this was true and had even visited Hitler at his secret retreat near Bariloche. It seems that Miguel Serrano knew much more than he ever wrote about.

This would be true of Savitri Devi as well. Because the two of them were guests of Otto Skorzeny and Hans-Ulrich Rudel, they would have been told about the Haunebu and the secret submarine bases and Hitler's escape to South America. In Serrano's case, he may have actually met with Hitler after the war in Argentina.

Had Miguel Serrano or Savitri Devi been given a ride in a secret craft? Perhaps the two of them were given a ride in a Haunebu or Vril craft while in Europe visiting Skorzeny or Rudel.

A typical ride in a Haunebu might look like this: You are told to go to a rural place near where you live that both you and the Haunebu pilot would be familiar with. You go to this place—and it can be broad daylight—and a Haunebu or a smaller Vril craft lands at the designated spot and you are invited on board. At this point you take a flight in the flying saucer, maybe for quite a distance. Then you are returned to the area you departed from and—Bob's your uncle—you've had a trip on a flying saucer!

Such a trip, easily arranged by Skorzeny or Rudel, would be impressive indeed. If your faith in the SS and Nazi ideals was faltering, then a ride in a Haunebu or Vril craft was just the thing to pick up your spirits. One wonders if Serrano might have been brought to the base in Neuschwabenland at some point in the 1960s as a reward for his efforts on behalf of the Nazis. He was sort of an international journalist that could be given some impressive rides on a flying saucer—even to Antarctica. Serrano hinted as much, but we will probably never know how much Serrano was shown of the extensive Nazi SS bases in Antarctica and South America.

With Serrano's special interest in Tibet and the Theosophical Masters one might think that he might have been flown to the secret SS base in Tibet where he would have been further impressed. This base was probably still functioning in the 1950s when Serrano was stationed in India. Was Savitri Devi also taken to the SS base in Tibet as a reward for her devotion to the Nazis? Such a trip would have been the time of her life and she had a love of travel. Also, where did Savitri Devi get all the money to do the extensive travelling that she did? Her books and booklets did not give her any real income. She relied on gifts and the largess of her many Nazi friends, who clearly funded her jaunts from Nazi sacred spot to Nazi sacred spot.

When looking at Miguel Serrano, Savitri Devi, Otto Skorzeny, Hans-Ulrich Rudel and others in light of the Haunebu, Vril, Andromeda and other craft, you have to assume that they knew much more during their time than we do now, and are just figuring out. With the many flying saucer sightings throughout the 1950s, 1960s, 1970s and 1980s, we know now that some—some—are of Haunebu and Vril craft across the skies of North and South America, Europe, Africa and Asia and elsewhere. The question is: what were they doing, where were they going, and who was inside these craft? Were they shuttling back and forth between the secret cities of the Black Sun? Just how extensive was the Third Power and the secret nests of SS rats that they were creating, the so-called secret cities of the Black Sun?

Grey Wolf: The Escape of Adolf Hitler

A curious book by two British researchers was published in Britain in 2011 called *Grey Wolf: The Escape of Adolf Hitler* by Gerrard Williams and Simon Dunstan.[4] This book was adapted as a drama documentary film directed and written by Gerrard Williams, produced by Magnus Peterson and released on DVD.

The book maintains that Adolf Hitler, whose code name was "Grey Wolf" in the Nazi hierarchy, did not die in his Berlin bunker in 1945 but escaped, along with Eva Braun and several other Nazi officials, to Argentina and lived six miles (10 kilometers) west of Bariloche in the Argentine Andes, near the Chilean border. According to the book, Hitler's escape was organized by Martin Bormann—who also fled to Argentina—and was aided and abetted by Juan Perón and his government. The American government was concerned that the Argentine government was working secretly with the Nazi regime during and after the war.

The book maintains that American intelligence agencies were aware Hitler was in Argentina and it reprints a number of

FBI documents that show that reports of Hitler in Argentina—apparently quite accurate—were made to American intelligence authorities. *Grey Wolf* also says that significant funds were taken from Germany to Argentina at the end of the war and that these funds were later stolen by Bormann. The book alleges that Hitler died in Argentina alone, poor and mentally ill in 1962, leaving behind a wife and at least one child.

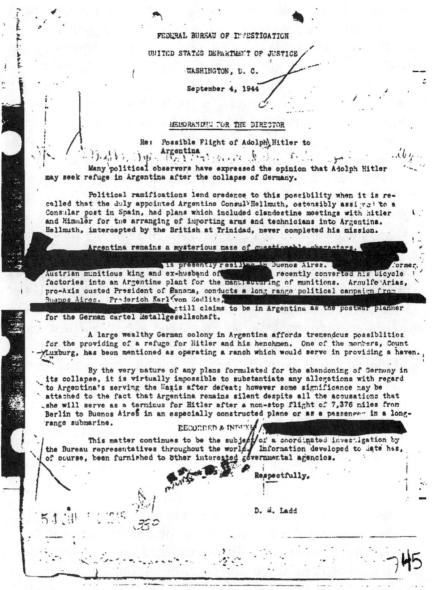

FEDERAL BUREAU OF INVESTIGATION

UNITED STATES DEPARTMENT OF JUSTICE

WASHINGTON, D. C.

September 4, 1944

MEMORANDUM FOR THE DIRECTOR

Re: Possible Flight of Adolph Hitler to Argentina

Many political observers have expressed the opinion that Adolph Hitler may seek refuge in Argentina after the collapse of Germany.

Political ramifications lend credence to this possibility when it is recalled that the duly appointed Argentine Consul Hellmuth, ostensibly assigned to a Consular post in Spain, had plans which included clandestine meetings with Hitler and Himmler for the arranging of importing arms and technicians into Argentina. Hellmuth, intercepted by the British at Trinidad, never completed his mission.

Argentina remains a mysterious maze of questionable characters.
is presently residing in Buenos Aires. former
Austrian munitious king and ex-husband of recently converted his bicycle factories into an Argentine plant for the manufacturing of munitions. Arnulfo Arias, pro-Axis ousted President of Panama, conducts a long range political campaign from Buenos Aires. Frederich Karl von Zedlitz,
still claims to be in Argentina as the postwar planner for the German cartel Metallgesellschaft.

A large wealthy German colony in Argentina affords tremendous possiblities for the providing of a refuge for Hitler and his henchmen. One of the members, Count Luxburg, has been mentioned as operating a ranch which would serve in providing a haven.

By the very nature of any plans formulated for the abandoning of Germany in its collapse, it is virtually impossible to substantiate any allegations with regard to Argentina's serving the Nazis after defeat; however some significance may be attached to the fact that Argentina remains silent despite all the accusations that she will serve as a terminus for Hitler after a non-stop flight of 7,376 miles from Berlin to Buenos Aires in an especially constructed plane or as a passenger in a long-range submarine.

RECORDED & INDEXED

This matter continues to be the subject of a coordinated investigation by the Bureau representatives throughout the world. Information developed to date has, of course, been furnished to other interested governmental agencies.

Respectfully,

D. M. Ladd

An FBI document dated Sept. 4, 1944 reporting that Hitler fled to Argentina.

149

The film that was made of the book had numerous production problems, bankruptcies and lawsuits that delayed its release, and while work on the film had begun in 2008, it was ultimately released in 2014. In 2013, as it was being completed, the film was hit by a further scandal when Abel Basti, an Argentine journalist, alleged that the *Grey Wolf* film and book plagiarized his work, and began legal action for compensation. Basti had previously published the book *Bariloche nazi-guía turística* (Spanish book titles typically start with the first word capitalized but following words not capitalized) in 2004 where he put forward claims about Adolf Hitler and Eva Braun living in the surroundings of Bariloche for many years after World War II, mentioning also the estate of Inalco as Hitler's refuge. The book also gives the same, familiar story (in Argentina at least) of Martin Bormann and Peron assisting Hitler, along with many other Nazis.

The proposed scenario is that a number of U-boats took certain Nazis and Nazi loot to Argentina, where the Nazis were supported by future president Juan Perón, who, with his wife "Evita," had been receiving money from the Nazis for some time. Perón would not become the President of Argentina until 1946, but was already a powerful political figure in Argentina during the war. Hitler allegedly arrived in Argentina on July 28, 1945, first staying at Hacienda San Ramón, east of San Carlos de Bariloche. Hitler then moved to a Bavarian-styled mansion at Inalco, a remote and barely accessible spot at the southwest end of Nahuel Huapi Lake, close to the Chilean border. Around 1954, Eva Braun left Hitler and moved to the town of Neuquén with their daughter, Ursula ('Uschi'). It is thought that there was a second daughter as well. Hitler died in February 1962 at a villa to the south of Nahuel Huapi Lake in a remote location.

An article in *The Guardian* on October 27, 2013 mentions *Grey Wolf* and Abel Basti:

> The notorious claim that Hitler escaped his Berlin bunker to live incognito in Argentina first gained popular currency in 1945, when Stalin spoke of it. Since then the idea has resurfaced occasionally, with alleged photographic and documentary evidence pored over by conspiracy theorists.

150

Now the theory that the German dictator followed his fellow Nazis Adolf Eichmann and Josef Mengele to South America is at the center of a fresh row.

The authors of the 2011 book *Grey Wolf: The Escape of Adolf Hitler*, which was made into a documentary film earlier this year, have been accused of plagiarism by a journalist in Argentina. Abel Basti claims his research has been unfairly used to substantiate claims made in the book. *Grey Wolf*, published by Sterling Publishing, based in New York, challenged the accepted view that the Führer shot himself in his Berlin bunker on 30 April 1945 and that Eva Braun also committed suicide by taking cyanide. Arguing that American intelligence officials turned a blind eye to Hitler's escape in return for access to Nazi war technology, Gerrard Williams and Simon Dunstan set out the case for a scenario almost too horrible to contemplate: that the Führer and Eva Braun made a home in the foothills of the Andes and had two daughters.

Hitler, they claim, escaped punishment and lived out his life in tranquility in Patagonia until his death in 1962 at the age of 73.

The publisher billed the book as the result "of five years of travelling and interviewing eyewitnesses and piecing together a mountain of evidence." Now Basti alleges that this is "a grossly misleading statement" and that Williams and Dunstan held on to evidence he had spent years putting together.

Williams, a British TV journalist who has worked for Reuters, the BBC and Sky News, and his co-author, Dunstan, firmly deny the claim.

"Basti did in no way invent the idea of Hitler being alive in Argentina," Williams told the *Observer*. "Books on the subject existed as far back as 1953 and 1987. I have never plagiarized anyone's work. Simon Dunstan, as the author of over 50 books on military history, hasn't either. We're both very aware of the law."

Williams travelled in 2007 to Argentina, where he acknowledges that he received help from Basti, along with other researchers and translators. Basti now claims that on seeing the book and hearing of the new film he realized that the work he had handed over to Williams for use on an earlier documentary film project had been plagiarized.

Basti says he signed a contract conferring all rights to his work to Williams's company in return for substantial payments to come. On this basis, he adds, he introduced Williams to two key witnesses in the case for Hitler's survival, a Jorge Colotto and Captain Manuel Monasterio.

Filming began in September 2008, but was cut short when financiers pulled out due to the worldwide financial crisis. Basti claims his contract was terminated and so asked for his research to be returned but says nothing was sent back.

Following publication of *Grey Wolf*, Basti says he was incensed to see that he had been quoted as regarding one photograph as proof of Hitler's survival. The book has also annoyed Ricardo D'Aloia, the editorial director of *Ambito Financerio*, the flagship newspaper of a group that had earlier published reports of Basti's research. D'Aloia is angered at the suggestion that he handed the authors video and potential evidence belonging to Basti.

Williams denies that he was introduced to key witnesses by Basti. He also denies D'Aloia's claims, which, he says, are "simply untrue." He adds that he "cannot see how quoting from taped interviews, thoroughly sourced to the company who made it implicitly clear that it was their material, is any sort of violation of copyright."

The claims about Hitler's life in exile in Argentina have been ridiculed by historian Guy Walters, who pronounced them "2,000% rubbish" when the book came out. "It's an absolute disgrace. There's no substance to it at all. It appeals to the deluded fantasies of conspiracy theorists and has no place whatsoever in historical research," he said.

This was what little publicity the book and movie *Grey Wolf* got. We learn that Basti seems to think, erroneously, that if one digs up a certain subject and publishes a book on it then they somehow have a copyright on the subject—not on pages and pages of verbatim text—but just the arcane subject itself. This is wrong, as subjects (or the titles of books or movies) cannot be copyrighted. Further, it is not a copyright violation to quote from a journalist's interview, as Williams points out. This argument was put to the test in London shortly after the millennium where certain authors tried to sue over the "Holy Blood, Holy Grail" concept about the Holy Grail. The London court struck down the claim that theories, concepts and discoveries can be copyrighted.

British historian Guy Walters speaks in superlatives describing the book as 2000% rubbish. Perhaps hogwash would have been a better term. He is a good example of the traditional historian, one who can dismiss talk of Hitler's escape as complete fantasy. I am sure that he would dismiss the existence of the Haunebu out of hand as well. Has he ever been to Bariloche, or even to Argentina? Has he actually done any research into the subject himself, or even read the book that he is dismissing as rubbish? I doubt it, because if he had, he would have seen some of the FBI documents presented in the book and concluded differently. If the FBI took this matter seriously would it really be 2000% rubbish?

> When Truman asked Stalin in 1945 whether Hitler was dead, Stalin replied bluntly, "No." As late as 1952, Eisenhower declared: "We have been unable to unearth one bit of tangible evidence of Hitler's death." What really happened?
> —*Grey Wolf*, Simon Dunstan and Gerrard Williams

The SS are still active in Argentina

Simon Dunstan and Gerrard Williams should be commended for their book as it is well researched. Simon Dunstan, as the author of over 50 books on military history, knows his stuff, unlike his critics. The authors have compiled extensive evidence— some recently declassified—such as the FBI reports. The recent discovery that the famous "Hitler's skull" in Moscow is that of a female helps them prove their case. The authors also suggest that the CIA had a possible involvement in Hitler's life in Patagonia— and with his two daughters.

In the documentary *Grey Wolf*, it is claimed that the Germans began investing heavily in Argentina during the war, including building a special hospital at Mar Chiquita near Cordoba in north-central Argentina at a cost of over 25 million dollars. They were also buying ranches in Patagonia, including an island on a lake near Bariloche that could only be reached by boat or seaplane. The book is much more heavily documented with detailed maps not included in the movie and more details about the escape from Germany. This subject is not dealt with at length in the movie.

The book says that Hitler and Braun flew out of Berlin on April 29, 1945 in a Junkers Ju 52 to the Nazi airbase in Tonder, Denmark. At the airbase in Tonder the party transferred to a different Junkers Ju 52 and flew a short distance back south to the long-range Luftwaffe base at Travemunde. Here they boarded a long-range Ju 252 for their flight to Reus, near Barcelona, Spain. Only 15 of the three-propeller Ju 252's were made, with production stopping in the fall of 1944 to conserve materials.

They then transferred to a Spanish air force Ju 52 which was to take the party to the Canary Islands. They took off from Reus and flew southwest to the airbase at Moron in southern Spain to refuel for the long trip to the secret Nazi airstrip and submarine base at Villa Winter on the eastern tip of Fuerteventura, the easternmost of the Canary Islands.

I write extensively about the Villa Winter and submarine base there in my book *Antarctica and the Secret Space Program*.[10] In

A Junkers Ju 52 photographed in February 1942.

A Junkers Ju 252 during the war.

that book I describe the Black Fleet of submarines that refused to surrender at the end of the war and left for various secret submarine bases around the world, especially in the Atlantic. Besides secret submarine bases in northern Norway, Greenland and Antarctica, the Germans had a secret U-boat base at Villa Winter at Fuerteventura. We will discuss the secret submarine base at Villa Winter in the next chapter, but suffice it to say that this isolated villa, complete with secret tunnels, is well known in Spain, Germany and the Canary Islands. Even Indiana Jones movies have featured this rock-cut submarine base that could hold three U-boats in berths at any one time.

Operation Seawolf, the authors of *Grey Wolf* maintain, was the U-boat operation to transport Hitler and Eva to South America. No, they were not going to live in Antarctica the rest of their lives, they were in fact going to a very Swiss Alps-type place where they would live much as they did when on vacation in Germany. It would be a pleasant alpine spot with glacial mountains, lakes and pine forests. Their young daughter, Ursula, was to stay in Spain and take a first class cabin on a passenger ship to Buenos Aires from there. She would meet them in Argentina.

After arriving at the isolated Villa Winter on Fuerteventura, Hitler and Braun would have spent the night at the heavily guarded estate. Supposedly Hitler and Braun then went down the stone-cut stairs in a tunnel from the Villa Winter to the secret submarine

A Junkers Ju 52 fitted with a minesweeper ring.

base on the coast beneath the villa. Here were three berths for U-boats. Hitler and Braun boarded U-518 and, after departing, make a 53-day voyage to Argentina.

German submarine U-518 was a Type IXC U-boat of the Nazi Germany's Kriegsmarine during World War II. She saw considerable success from her launch on 11 February 1942. This submarine made several voyages to the Azores, Canary Islands and Brazil during the war. According to historians, her last foray began when she departed Kristiansand, Norway on March 12, 1945 and was supposedly sunk northwest of the Azores on April 24 by destructive explosions called hedgehog rounds from the USS *Carter* and the USS *Neal A. Scott*. The submarine is listed as sunk with no survivors. Yet, we have U-518 in use at the Canary Islands base in July of 1945. As I explain in my book *Antarctica and the Secret Space Program*, a number of U-boats that are listed as sunk were still active after the war, and in some cases, there were actually two U-boats with the same number. So, it would appear that U-518 was not sunk near the Azores, where there was no real evidence the submarine had been sunk except that it became inactive and was no longer in the area. It seems to have slipped silently away, and then proceeded to the Canary Islands. Dunston and Williams discuss this and say that no wreckage of the sub ever came to the surface; they mention that two other submarines were supposedly sunk in the same action, the U-880 and the U-1235. These may have been the two other submarines at the secret U-boat base at Villa Winter.

The authors of *Grey Wolf* say that U-518 arrived at the Argentine coast at Necochea, a small town just south of the major port of Mar del Plata, on July 28, 1945. A day or so later the group flew on July 30 out of Necochea on an Argentine air force Curtiss biplane to the city of Neuquén in northern Patagonia where the plane was refueled. The group then flew south to the Estancia San Ramon ranch on the outskirts of the Andean resort town of Bariloche where Hitler and Braun began their exile in South America. The book does not mention any stop in Antarctica on the way to Argentina, but simply says that the voyage was 53 days.

They stayed at the Estancia San Ramon for about nine months or so and were joined by their daughter, Ursula ('Ushi'). The book maintains that Hitler and Eva Braun faked their deaths a second time while in the Bariloche area by having everyone who had seen them in 1945 or early 1946 told that Hitler and Eva had died in a car crash near Bariloche.

The group then moved to the hotel/hospital called the Hotel Viena near the northern town of Cordoba on the shore of a lake called Mar Chiquita. At this exclusive German hotel/hospital— newly built in 1943 at a cost of 25 million dollars—they had their own dinnerware with the initials A.H., and nearly the entire hotel staff were Germans from Buenos Aires. After a year or so in Mar Chiquita, Hitler and Eva moved to the island on the lake near Bariloche and lived on an estate called Inalco. It was now 1947. Apparently Hitler and Braun then had a second daughter whose name is unknown.

In *Grey Wolf*, the authors say that Martin Bormann visited Hitler at his estate and told him that the Nazis had amassed a huge fortune that could be used to finance the resurgence of the Third Reich, mainly in South America, and the money could be used to bribe and influence politicians and journalists.

In 1955 Peron was deposed in a military coup and the situation for Nazis in Argentina began to change. Bormann was to keep Hitler and the other Nazis safe and they still had plenty of money.

157

But Bormann came to Hitler at his estate of Inalco and told him that their plans must be put on hold. At about this time Eva Braun decided to leave Hitler and take their two daughters with her. No one would be looking for a woman on her own with two daughters. The three of them moved to the city of Neuquén, the capital of Neuquén Province, north of Bariloche. Supposedly Eva raised her daughters here through the 1960s and eventually moved to Buenos Aires without them.

Hitler moved to a place called La Clara, an isolated villa not

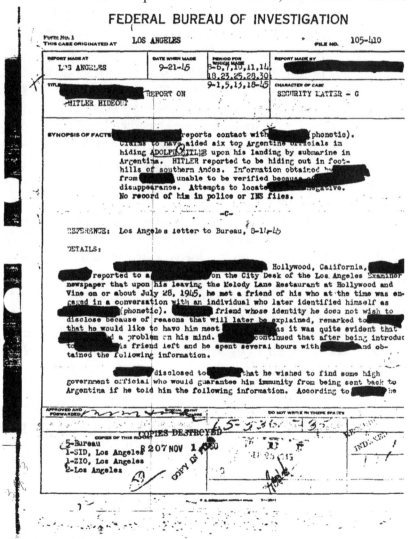

The second page of an FBI document from Sept. 21, 1945 reporting that Hitler fled to Argentina.

far from Inalco to the south, away from Nahuel Huapi Lake. According to *Grey Wolf*, Hitler remained here until his death in February 1962.

Bormann now believed that any hope of there being any real political movement or uprising that could be implemented to bring the Nazis back to power in Argentina, Europe or elsewhere was lost. The war had been over for ten years now and Nazi survivors

105-410

was one of four men who met HITLER and his party when they landed from two submarines in Argentina approximately two and one-half weeks after the fall of Berlin. ███████ continued that the first sub came close to shore about 11:00 p.m. after it had been signaled that it was safe to land and a doctor and several men disembarked. Approximately two hours later the second sub came ashore and HITLER, two women, another doctor, and several more men, making the whole party arriving by submarines approximately 50, were aboard. By pre-arranged plan with six top Argentine officials, pack horses were waiting for the group and by daylight all supplies were loaded on the horses and an all-day trip inland toward the foothills of the southern Andes was started. At dusk the party arrived at the ranch where HITLER and his party, according to ███████, are now in hiding. ███████ most specifically explained that the subs landed along the tip of the Valdez Peninsula along the southern tip of Argentina in the gulf of San Matias. ███████ told ███████ that there are several tiny villages in this area where members of HITLER's party would eventually stay with German families. He named the towns as San Antonio, Videma, Neuquen, Muster, Carmena, and Rason.

███████ maintains that he can name the six Argentine officials and also the names of the three other men who helped HITLER inland to his hiding place. ███████ explained that he was given $15,000 for helping in the deal. ███████ explained to ███████ that he was hiding out in the United States now so that he could later tell how he got out of Argentina. He stated to ███████ that he would tell his story to the United States officials after HITLER's capture so that they might keep him from having to return to Argentina. He further explained to ███████ that the matter was weighing on his mind and that he did not wish to be mixed up in the business any further.

According to ███████, HITLER is suffering from asthma and ulcers, has shaved off his mustache and has a long "but" on his upper lip.

███████ gave the following directions to ███████ "If you will go to a hotel in San Antonio, Argentina, I will arrange for a man to meet you there and locate the ranch where HITLER is. It is heavily guarded, of course, and you will be risking your life to go there. If you do go to Argentina, place an ad in the Examiner stating, ███████ call Hempstead 8458,' and I know that you are on the way to San Antonio."

The above information was given to ███████, reporter on the Los Angeles Examiner on July 29, 1945.

The writer contacted ███████ in an attempt to locate ███████ in order that he might be vigorously interviewed in detail concerning the above store. ███████ reiterated the information set out above, adding that the friend to whom ███████ was talking in front of the Melody Lane Restaurant was a friend of his by the name of "JACK," last name unknown, but that since the introduction he has had further conversation with "JACK" and "JACK" advised him that while he was eating his lunch at the Melody Lane Restaurant ███████ sat at his table

-2-

3

Page three of an FBI document dated Sept. 21, 1945 reporting that Hitler fled to Argentina.

in Argentina, Chile, Paraguay and Brazil were getting older, and though they still had money, they were no longer a significant power on the world stage.

At La Clara, Hitler received a visit from Bormann, arriving from Chile. Having been depressed for a year or so, Hitler became a bit optimistic. Yet, nothing was really to happen and Hitler got older and sicker, returning to his passion of painting, and living with only one bodyguard and a doctor. By 1961 he was deteriorating both mentally and physically. In the film of *Grey Wolf*, Hitler is seen weeping alone in his lonely room, a miserable failure who has let the people who adored him down. In the film, he is shown being given drugs to calm and suppress him. Eventually, Hitler dies of his illness in February of 1962. Hitler is cremated and his ashes scattered in a graveyard.

Hitler's health was complicated by a wooden splinter deep in his nose that remained from the failed assassination attempt on him on July 20, 1944. On this date, one year or so before Hitler's arrival in Argentina, Claus von Stauffenberg and other conspirators attempted to assassinate Hitler inside his Wolf's Lair field headquarters near Rastenburg, East Prussia. The name Operation Valkyrie—originally referring to part of the conspiracy—has become associated with the entire event. The apparent aim of the assassination attempt was to wrest political control of Germany and its armed forces from the Nazi Party—including the SS—and to make peace with the Western Allies as soon as possible. Because the heavy wooden leg of a table shielded Hitler and others from a bomb in a briefcase, the failed assassination attempt meant that the SS was still in control of all military intelligence—and all of the submarines and aircraft at its command—at the end of the war, a war in which the SS never surrendered.

Meanwhile, in a curious part at the end of the *Grey Wolf* film, there is a scene that has a British investigator in 2008 threatened in a telephone call if he kept pursuing the Hitler in Argentina story. Another person, one who had seen Hitler in 1946, a woman, is also warned by a disturbing phone call, also in 2008, that she should stop speaking about this story and that the SS and Gestapo were still active (in Argentina) and that Eva Braun was still alive. The film then shows Eva entering a hospital in Buenos Aires in 2008 and hints that she died there.

The scenario laid out in *Grey Wolf* is detailed and credible. The authors completely steer away from the subjects of more U-boats patrolling the Atlantic, an Antarctic base, or the Haunebu or other exotic craft. Until recently, such topics were largely ignored by historians, though they crept into popular culture through books and movies.

At least 30,000 Nazis escaped to South America after the war and the SS was essentially in control of this diaspora. As in the United States with Operation Paperclip, Argentina also welcomed displaced German scientists such as Kurt Tank and Ronald Richter. The Argentine consulate in Barcelona was said to have given thousands of false passports to fleeing Nazi war criminals and collaborationists.

When Perón became president of Argentina on June 4, 1946, his two stated goals were social justice and economic independence. Perón instructed his economic advisers to develop a five-year plan with the goals of increasing workers' pay, achieving full employment, stimulating industrial growth of over 40% while diversifying the agricultural sector and greatly improving transportation, communication, energy and social infrastructure. Where would he get the money for all these projects?

Perón allowed Nazis into the country in hopes of acquiring

The Fábrica Militar de Aviones de Córdoba factory in Argentina, 1950.

advanced German technology developed during the war. He also wanted German investment in Argentine automobile and plane manufacturing, and was eager for Argentina to have its own automobile. Perón got all of this, including money from Bormann and German companies like Mercedes Benz, Porsche, and Volkswagen setting up factories in Cordoba, the Detroit of Argentina. Perón even had a special car designed called the "Justicialista" which was produced by the government of Argentina via IAME (Industrias Aeronáuticas y Mecánicas del Estado) from 1954 to 1955. It was an early attempt to form an Argentine automotive industry and used a front-engine, front-wheel-drive layout with a two-stroke two-cylinder engine derived from a German DKW design with a conventional metal body. Due to the insistence of Perón to have a sports car version, a fiberglass two-seat version was available as a coupé or roadster, powered by a 1.5-liter air-cooled Porsche flat-four engine and a Porsche four-speed gearbox driving the front wheels. Both cars featured a special logo-shield of the Perónist party on the rear bumper. They were manufactured in Cordoba but they were discontinued after Perón was overthrown in a coup in 1955. But these projects helped bring more cars, trucks and roads to a large and undeveloped country.

The final hanging thread of *Grey Wolf* is what happened to all the money that Martin Bormann supposedly "stole" from Hitler and the SS? The authors of *Grey Wolf* largely seem to think that Nazi activity basically stopped in Argentina after Hitler's death in 1962, though they feature the aforementioned two phone calls in the film that essentially claim that the SS is still operating in Argentina.

Yet, what of all the UFO activity in Argentina and elsewhere? As we will see in the next chapter, UFO activity in South America increased dramatically in the 1950s and continues to this day. While not all of this UFO activity can necessarily be attributed to postwar Nazis flying Haunebu and Vril craft around Argentina and elsewhere, I think it is safe to say that certainly some of these sightings are of German-made craft. And, some of these craft were probably made in Argentina—at one of the secret cities of the SS. Is this where all of the money went that Martin Bormann had control of? It had to go somewhere.

Chapter 5

Secret Cities of the Black Sun

Everybody's talking and no one says a word
Everybody's making love and no one really cares
There's Nazis in the bathroom just below the stairs
Always something happening and nothing going on…
Everybody's flying and never touch the sky
There's UFO's over New York and I ain't too surprised
Nobody told me there'd be days like these
Strange days indeed.
—John Lennon, *Nobody Told Me*

In the last chapter we mentioned the Villa Winter in the Canary Islands and how Hitler and Braun were alleged to have flown from Spain to the private airstrip at Villa Winter and then take a U-boat to Argentina in July of 1945. It should be said that Hitler's possible escape from Germany to Argentina is not really very important to the greater theme of this book, the subject of the Haunebu and its use after the war. As I mentioned in the last chapter, the authors of *Grey Wolf* never mention the subjects of Antarctica, Neuschwabenland, or German flying saucers. In the book they paint Hitler as a weak and sickly man who did little else but read books at his mountain retreat near Bariloche. Then he died and all the money that Martin Bormann had control of just disappeared.

Hitler is clearly secondary to the secret activities after the war involving U-boats, flying saucers and secret bases around the world and it does not matter whether Hitler died in Berlin or not. Despite Miguel Serrano and Savitri Devi's beliefs that Hitler was some sort of superman-avatar as in Serrano's bizarre astral journey to visit his hero at the secret base in Antarctica, Hitler did little to further the goals of the SS—whatever they were.

I suppose Otto Skorzeny and Martin Bormann knew better. That Hitler had been to the secret base in Antarctica had already been published in the 1947 book *Hitler es vivo*.[19] Was this one of the volumes in Hitler's library at his estate? Remember, this was a book that had Hitler living in Antarctica in 1945-46.

The vast amount of Nazi money was still controlled by Martin Bormann, much of it gold brought by submarine to Antarctica and Argentina. In 1947, it was Bormann's goal to fund the secret SS bases around the world. Where were they? We have already discussed the remarkable SS airbase in Tibet. Let us take a closer look at the secret submarine base in the Canary Islands that turned out to be quite important immediately after the war.

The Secret Submarine Base in the Canary Islands

The secret base on Fuerteventura is a fascinating story and gives us good insight into what a secret submarine base might be like—something straight out of a Hollywood movie. The island of Fuerteventura is a large, rocky island in the eastern part of the Canary Islands, and is the closest island to the African coast. This part of the coast was the Spanish Sahara, an area that welcomed many ex-Nazis after the war. It was even rumored that Adolf Hitler was living in the Spanish Sahara (when it was still a colony of Spain).

Says Henry Stevens in *Dark Star*:

> There is no longer any doubt that a secret German base existed on Fuerteventura during the Second World War. One is reminded of the statement by Admiral Karl Dönitz about how the German Navy knew all the ocean's hiding places. Well, this is certainly one of them—a secret base in plain sight.
>
> This base has been mentioned by me in both my earlier books. Others have discussed this base, writing in the German language. For some reason general knowledge of this base has not penetrated the consciousness of the English-speaking world, probably because of the language

barrier. I say this because the real description of the base is in German that has never been translated into English.

Fuerteventura is a resort spot but an out-of-the-way one. From discussions I have heard there is a small tourist town there that includes a bar. Rumors of the secret German base have always been discussed and questions asked about it, mostly by tourists. Evidently, everyone or most everyone on the island denies these rumors in public but affirm them in private. But everyone, even you dear reader, have heard of this base in a roundabout way. This was the secret U-boat bunker featured in the George Lucas/ Steven Spielberg film *Raiders of the Lost Ark*. In that film a U-boat sailed right inside the island, using a tunnel, into a huge cavern that was perhaps an ancient volcanic blister. There, the Nazi bad guys had erected submarine support equipment, turntables for U-boats, along with supporting manpower.

In 1971 *Stern* magazine published an article about the father of this base and the base itself in an article titled (and translated to English) "The Fantastic History of Don Gustavos, His Secret House and the U-Boat Base." According to the rumors and the facts on the ground, the father of this secret base was a German General, Gustav Winter. General Winter built a large, white villa on the high point of this island. The rumor was that a staircase descended down into the bowels of the island, connecting to this secret base.[46]

Stevens says that the German researchers Heiner Gehring and Karl Heinz-Zunneck did some research on General Winter, which revealed a few things about this secret U-boat base. Winter was born in 1893 in the Black Forest and was trained as an engineer. He died in 1971, a few months before the *Stern* article was published. General Winter performed some outstanding service for the Germans, and for that service he was granted land on the southern peninsula of Fuerteventura. During World War II, General Winter was the driving force behind a project to build a secret U-boat bunker on the island. He was also an agent for the Abwehr, one of the German spy services. The major feature of this bunker was a huge natural cavity that was connected to the sea by tunnels bored into the solid rock.

Ventilation shafts were dug upward towards what was now the large Villa Winter, which also included an airport. This facility functioned as a military base during the war and was heavily armed and guarded. After the war parts of the facility, with the exception of the airport, were destroyed with explosives. Today part of this large site belongs to a nature park but part of the property is still privately owned.

Stevens tells us that the best information comes not from *Stern*, but from another German magazine, *Nugget*. *Nugget* was a German-language magazine of treasure hunters. Below the title appeared the words (in English) Gold, Minerals, Treasure Seeking,

The Villa Winter on Fuerteventura. A tunnel led to a secret U-boat base.

Adventure. Stevens says that a two-part article was published in the July and August 1984 editions titled: "The U-Boat Bunker of Fuerteventura" (Parts One and Two). The article describes the adventures of two Spaniards and one Austrian who use a yacht and scuba gear to anchor off of the coast where the secret entrance to the submarine base was located and dive to the secret bunker.

The two divers (the Spaniards) slipped into the water on the ocean side of the yacht, so as not to be noticed by anyone on shore, and then swam underwater to the coastline that was essentially a volcanic cliff. A crack in the rock of this giant cliff led to a large natural cavern deeper inside the island. The Germans had evidently widened this tunnel-crack in the rock and reshaped the interior cavern, which was partly underwater and partially out of the water, and made three different submarine pens plus other rooms, carved out of the solid rock.

The two divers swam through the tunnel and then climbed up a metal ladder into the underground base. To their astonishment they saw that there were three submarine pens and two of them had submarines docked in the dark silence. They attempted to open the hatch of the first submarine and failed. They went to the second submarine and attempted to open the hatch. This time they

167

succeeded.

Inside the submarine everything was neat and tidy. There was no sign that there had been any trouble or that the crew had left in haste. There were charts of the South American coast and other areas. Six points were marked on the coast of South America on one of the maps by a T with a circle around it. Between the various maps were newspaper clippings.

They left the piles of maps as they found them and exited the U-boat. They noted the numbers of both U-boats (but do not give them in the story) and later discovered that both of these U-boats were officially listed as sunk. They put their scuba gear back on and swam out through the underwater tunnel and back to the waiting yacht. They carefully climbed on board from the side that faced the ocean and not the island. The group then departed.

The article does not end there. The group continued to hang out in the Canary Islands, and the author says that they met several people who could tell them about the secret German U-boat base. They were told by one informant named "Charlie" that the base was fairly well known in the Canary Islands and that General Winter blew up the secret stairway that led to the U-boat pens in 1945.

The submarine bunker was still active until about 1950 they

A view looking down at the Villa Winter on Fuerteventura.

were told, though only accessible from the ocean. They were told that the base was known to the Americans, who allowed it to be used, the informant said, to bring Project Paperclip Nazi scientists and officers secretly to the United States. They were told that there were originally three subs that came and went from the base but one of them was sunk near Florida—another missing sub, one that was probably on the list above. These submarines, and the base, were being operated by the SS during and after the war. The crews of the submarines were still living in the Canary Islands in many cases, though some had gone back to Germany. Also, in the book *Grey Wolf,*[4] the authors maintain that the base was used by U-518, a U-boat that was supposedly sunk on 22 April 1945. This U-boat was still active, as were two other U-boats supposedly sunk at the same time, the U-880 and the U-1235. All three of these subs may have used the Villa Winter submarine base; all three were believed sunk, so no one was looking for these "missing" submarines.

Says Stevens:

> Here in the Canary Islands was an SS base that operated after World War Two. In fact, two U-boats were on-call and continued to be on-call for some time even after their last mission. We know this because the informant, "Charlie," includes the detail that the crew lived in the Canary Islands.
>
> This base functioned at least up until 1950 and with the knowledge and consent of the Americans. This would be the intelligence services of the USA, probably the CIA. We will get into this relationship later but we begin to see that this relationship between the surviving SS organizations, which became the "Nazi International" or The Third Power, was not always adversarial.
>
> Knowledge of this base and its continued existence after the war brings up claims of other facilities, laboratories, being operated by the SS after the war. Yes, there are those that say this is so and was done, not over the horizon in some never-never land, but right in Europe itself![46]

Stevens goes on to tell the story of a "Mrs. Maria W." who was aware of secret facilities in Western Germany after the war, including a secret SS research facility in the Jonas Valley. She described a small train track running to Bienstein (presumably also from the area of the Jonas Valley) that ran into a mountain. The Russians found the train track but did not ever follow the track fully into the mountain. This was the entrance to a German underground laboratory. For whatever reason, the Russians did not investigate the German facility there. "Mrs. W" says this was a "Fusionsanlage" an atomic fusion facility. Not fission but fusion, as in a hydrogen bomb! And importantly, the woman says this facility operated until 1952! At that time it was supposedly shut down but remains ready to fully resume operation.

So, here we have an SS laboratory operating for seven years after the war in West Germany. It seems that the SS were able to maintain a few secret facilities in West Germany, at least. All areas of East Germany were probably pretty well scoured by the Russians, who were the real enemy during the war. Britain and the West were adversaries, but many in the Third Reich had hoped until the end that they could negotiate a truce with Britain to allow them to focus on Russia and holding on to territory they had taken in the Ukraine. Nazi officers rushed to surrender to the British or Americans, knowing that they would be better treated.

The U-218 departs the port of Kiel in 1941.

Indeed, Operation Paperclip began immediately after the war and Nazi scientists like Wernher von Braun were brought to the United States. Others, apparently, came via U-boat from the Canary Islands to Florida in secret CIA missions in connection with the SS. So, we can see why the FBI was collecting its own files on escaped Nazis and Hitler fleeing to Argentina in a U-boat. The FBI apparently suspected that the CIA was bringing former Nazis and SS officers to the United States. There are even reports in the FBI files of Hitler being seen at certain cities in the Eastern US.

Well, the FBI was right, the CIA was doing all of that and more. The CIA attempted to work with the remnants of the SS as much as they could without alerting the FBI, the military rank and file, or the public. Exactly how much the CIA knew about secret SS operations in the Atlantic we will probably never know.

There may have been as many as a hundred black U-boats in the Ghost Fleet that refused to surrender on the hour that Admiral Dönitz had proclaimed, though the figure is likely much lower— typically put at 35 for the submarines that assembled in Greenland at the end of the war, to be discussed shortly. The submarine base at Fuerteventura would have been well used as a safe intermediary base to go from such areas as Norway, Spain, and the remote Spanish Sahara (south of Morocco and controlled by Spain until it was ceded to Morocco in 1975) to areas further south such as Antarctica, South America, South Africa and beyond.

Stevens says that the existence of the base at Fuerteventura gives us a glimpse of what the other secret U-boat bases were like in Greenland and Antarctica:

> But the thing which should most concern us is the description of the base and its almost certainty of actual existence. If you accept the description of the base as a huge, hollow, natural volcanic bubble, connected to and accessible to the sea by tunnels, being full of machinery necessary to run a U-boat bunker, almost in plain sight, overflown and monitored on the surface by Allied aircraft and ships, then the proposition must be taken seriously that the Germans also had similar bases in areas where they would have been able to construct them unhindered. The

fact is that the German base at Fuerteventura alone makes the case for the strong possibility if not the probability of similar bases in the Arctic and Antarctic.

Finally, Stevens tells us that the "outstanding service" performed for the German nation by General Gustav Winter, an engineer, for which he was given land on Fuerteventura and oversight of the base on the island was a curious invention. Before the war Winter worked on research that used strong magnetic fields to influence the compasses of ships and U-boats. This seems to have resulted in the Magnetofunk device that sends out strong magnetic signals that confuse the compasses and other parts of aircraft and ships. This Magnetofunk device was also used at the secret U-boat base in northeastern Greenland, another Arctic Base known as Point 103, said to be in the Canadian Arctic, apparently around Baffin Island, and probably the secret Antarctic base known as Point 211. The function of the Magnetofunk device was to divert enemy aircraft that were seeking to find these secret bases in order to destroy them. The mysterious Foo Fighters over German skies at the end of the war apparently had a similar function: to interfere with the electrical systems and navigation of the Allied bombers.

NEW NAZI WEAPON: MYSTERIOUS BALLS WHICH FLOAT IN AIR

PARIS, Dec. 13 (AP).—As the Allied armies ground out new gains on the western front today, the Germans were disclosed to have thrown a new "device" into the war—mysterious silvery balls which float in the air.

Pilots reported seeing these objects, both individually and in clusters, during forays over the Reich.

(The purpose of the floaters was not immediately evident. It is possible that they represent a new antiaircraft defense instrument or weapon.

(This dispatch was heavily censored at Supreme Headquarters.)

A 1944 newsclipping about Foo Fighters.

The Secret Nazi Arctic U-Boat Bases

While a Nazi base in Antarctica is now fairly well known, even in popular culture, it is not so well known that there were two or more secret Nazi submarine bases in the Arctic. One was in northeast Greenland called "Beaver Dam" and one was in the Canadian Arctic somewhere on or near Baffin Island, called Point 103. Point 103 was supposedly seen as a "Blue Island" by Canadian air patrols. Both of these bases were in contact with an airbase at Banak in the very northern part of Norway. The airfield was first built with triangular runways in 1938 and was taken over by the Luftwaffe in 1940, who expanded it and laid down two wooden runways.

Banak was taken over by the Royal Norwegian Air Force in 1945, but abandoned in 1952. At Banak, long-range aircraft were based that were operated by the SS, and some of these planes were said to have the insignia of the SS cult of the Black Sun on them, rather than the normal German Cross of the Luftwaffe. It was from this airfield that long-range flights to Japan took off, as well as flights to the bases in Greenland and the Canadian Arctic. The Norwegian arctic territories were also controlled by the Germans, and ultimately by the SS—because they were the only military intelligence at the end of the war.

Henry Stevens says that the first mention of the secret Greenland base called Beaver Dam was in an article in the German magazine *Mensch und Schicksal* written by Claude Schweikhart in 1952, talking about a Nazi-hunting group called the Emerich-Team, the first investigative body ever to look into the esoteric Reich and the "Third Power." In the article Schweikhart makes mention of a "Laboratory of Death" in Greenland, apparently the base called Beaver Dam, and in another statement concerning flying U-boats the magazine said:

> Interestingly enough the Emerich-Team recorded in such documents, obviously working together with the Third Power-hunting Yugoslavian radio news service, the existence of a "Laboratory of Death" somewhere in Greenland which was designated a city of technical satanism and in 1951 is said to have tested a flying undersea boat.[46]

173

A flying undersea boat is a flying submarine. As the war drew to a close and Germany was losing, what was left of the German navy—largely submarines—essentially withdrew to occupied Norway. Norway contained a number of U-boat bases and with its many fjords was an ideal place to hide submarines and even conventional boats. Says Stevens:

> The central Norwegian coast is fjord country. These fjords harbored the German Navy during the war. At first it was the refuge of capital ships such as the Bismarck and Scharnhorst but later these facilities took on an even greater importance as U-boat bunkers in France were bombed and eventually overrun. The U-boat facilities in Norway became the home for the U-boat fleet and the U-boat fleet became the real German Navy.
>
> It was from this area that U-boats were sent out in the final days of the 3rd Reich on all manner of desperate missions. Until the very end, U-boats apparently awaited Nazi bigwigs who never showed up for their journeys of escape. Later, we will recount the mission of U-234 as well as the U-boats of the legendary "Last Battalion" whose point of departure was exactly these waters.
>
> But for now we must return to the springboard for the Nazi postwar world, the airbases of Northern Norway

A U-boat at sea.

from which the Germans explored and watched over their holding in the Arctic. From here special long distance aircraft, including prototype aircraft salvaged from the jaws of Germany's collapse, were operational. Also, something new was in the air, a new identity was forming. Not only Wilhelm Landig tells us this in 1971, German engineer Claude Schweikhart (sometimes aka Erich Halk) tells this to us for the first time much earlier. In the August, 1952 issue of the magazine *Mensch und Schicksal*, he states that often in the last days of the war, in areas of highest priority for the Germans, the symbol of the "Schwarzen Sonne" (Black Sun) was in use in portions of the German armed air defense forces. These "heretics" within the SS, as Schweikhart calls them, above all used the air base at Banak in Northern Norway as their base for long distance flights over the inner Arctic using the most modern long distance aircraft which they had obtained after these aircraft were designated as "failed constructions" by the official Luftwaffe. Speaking specifically of this base at Banak, Schweikhart says: "If they likewise used a special symbol, it is uncertain, but possible."

This "failed construction" methodology for procuring long-distance aircraft is repeated by Landig and said by him to include flying discs. These aircraft were the property of a sect within the SS and were re-branded with the Black Sun symbol according to Landig.

Meanwhile, back in the highland plateau region, the Germans there evidently had ideas of carrying forth the war from Norway or at least prolonging the war from there. In fact, the last known German unit to surrender on the European mainland did so at Haudegen, Norway, on September 4, 1945. This date lacks four days from being four months after Admiral Doenitz officially surrendered the three German fighting divisions of the German military, the Army, the Navy and the Air Force. Please note that the SS apparently never officially surrendered.[46]

Stevens tells us that most of the information we have about

these secret Arctic bases manned by the SS comes from Wilhelm Landig's book published in the 1970s called *Goetzen Gegen Thule* (*Idols Against Thule*), the first book in a trilogy. Early in this novel three SS officers who are stationed in northern Norway leave their base in a special long-distance aircraft bound for a secret German postwar base somewhere in the Arctic, "Stuetzpunkt 103," or as it is commonly called, Point 103. Says Stevens:

> Wihelm Landig is just about our only source for the secret base he calls Point 103. Landig uses every setting, every situation, to unveil something that happened during the war or some aspect of thinking or technology. When his three characters, the three German soldiers, land at Point 103 they are told of the "V-7", a German flying disc that has been brought there. They are told its developmental history and means of operation. They are likewise informed of other prototype aircraft that have been brought to this base and housed there as well as the extent and purpose of Point 103 itself.
>
> Point 103 is set by Landig to be in the Canadian Arctic. We know that the Germans had a base as far West as Cape Chidley which is east of Hudson Bay but Landig's locating this particular base in the Canadian Arctic rather than farther east may be literary license, we simply do not know. It is set on an island in a semi-circular basin surrounding Point 103. Tunnels were made in the rock and served as living areas and storage as was done by the Germans at many other of their secret facilities, Nordhausen, Jonastal, Der Riese, Zement, and so on. Since aircraft were flown into this base, it must have had runway facilities. Hangars were blasted out of the solid rock mountains surrounding the base and it was in these cave-hangars that experimental aircraft of all sorts were stored. This base contained workshops, research facilities, housing, cooking facilities, offices, corridors, and assembly halls. Hundreds of German soldiers inhabited this base on an ongoing basis. It functioned as a listening station, air base, weather station and most probably a U-boat facility.

Stevens says that Point 103 was connected to the outside world with special radio transmissions that were sent to a special radio base in the Northern Andes. This may be the radio station that the Germans allegedly built at Machu Picchu near Cuzco, Peru. The Germans also built the beer brewery in Cuzco during this same time and that brewery is one of the few industries that exist in Cuzco. There may well have been another secret radio station—and perhaps Haunebu base—in the mountains of Venezuela somewhere.

A great deal of Nazi activity, as well as UFO and Men in Black activity, seems have gone down in Venezuela over the years, especially in the 1950s and 60s. Some of this activity may have stemmed from operations during the war, including a secret radio transmitter and receiver. Still, we don't where it was, and it might have been in Peru, Ecuador or Colombia as well.

Stevens also says that flying disks were used to fly to various parts of the world to remain in contact with Nazi elements that remained after the war. Says Stevens:

> Writer O. Bergmann states that Germans of the Arctic groups were in communication with other postwar German groups via in intermediate position in the Northern Andes. Each group had its own code-name. The group in the Northern Andes used the designation "Atlantis," for instance. He states that other postwar groups survived in bases in inner Asia, as well as Africa and that these along with the Arctic base formed a unit while another three group unit was formed by a group in Greenland, somewhere near Australia and somewhere in the Pacific.

With this we have a pretty good look at the worldwide network of SS bases—the secret cities of the Black Sun—and their far-flung locations. He mentions inner Asia and this would be the Haunebu base in western Tibet that was discussed earlier in the book. Bergmann then mentions Africa as a location for one or more SS bases. I would guess that these bases would be located in the Spanish Sahara and on the northern coast of Namibia, near the border with Angola. Both of these areas were known for Nazi activity and Hitler was said to be living in the Spanish Sahara in one early account. This may have been a ruse to disguise the fact that Hitler had been in the Canary Islands, which are very nearby, and also a Spanish possession.

Bergmann mentions Greenland bases and then says there was an SS base near Australia and one somewhere in the Pacific. A base near Australia might be a location such as the highlands in the interior of New Guinea, a largely unexplored island—one of the largest in the world—that is politically split between two countries and has very little infrastructure. A Haunebu base could easily exist here on the top of a mountain in a remote area.

Such a base would be completely self sufficient and isolated from the surrounding world. Materials such as cement and plastics

A U-boat in its protected slip.

would be brought in aboard a Haunebu along with tools such as saws and shovels. Basic buildings would be built and a generator and radio tower would be installed. A rainwater entrapment system would be created. A cave system might be utilized as well. A base for a Haunebu need not be so elaborate as a base for a U-boat, or as difficult to construct.

A similar base might exist on one of the islands in the Solomon Islands, or perhaps on one of the islands of Vanuatu. Odd UFO activity has been seen in all of these locations, particularly the Solomon Islands. It seems that there was some UFO activity in Japan immediately after the war and this activity may have been Haunebu craft flying out of these Pacific bases, or the one in Tibet, to bring SS operatives out of Japan after the war. Some of the strange UFO activity around Australia and New Zealand after the war may have come from these bases as well.

Also, it is worth noting that Indonesia—a vast archipelago of many mountainous islands and steep coasts—was occupied by the Japanese during the war. It is known that U-boats made journeys from the Atlantic to the Indian Ocean, past Indonesia to Japan, while Japanese submarines made this same trip to the Atlantic Ocean and back. Did some of these U-boats and Japanese submarines stop at a secret submarine base somewhere in Sumatra, Sulawesi or elsewhere? Could such a secret submarine base have functioned after the war?

Back in the Arctic, Stevens says that some of the craft kept at Point 103 were part of a special SS squadron, and that these craft began using the Black Sun symbol as their insignia rather than standard German insignias such as the Iron Cross. Says Stevens:

> It seems that all branches of the German military were represented at Point 103, including the SS. The particular sub-group or inner circle of the SS present at Point 103 lent their esoteric symbol, the Black Sun, to the base as a whole. On their aircraft the standard German insignias were removed and in their place the Black Sun sign was displayed. This took the form of a deep, dark-red circle. It is quite obvious that the thinking behind this symbol came from the Thule Society and their belief in a kind

of Nordic Atlantis in the far north from which sprung the Aryan peoples, particularly the ancient Germanics. The Black Sun symbol reached its zenith at Point 103. It was adopted by some mystery U-boats of German origin sighted after the war but it was apparently not used in Neuschwabenland, Landig's Point 211. The flying discs and possibly aircraft flying from Point 211 used the old German military insignias.[46]

Stevens mentions how during the war, Wilhelm Landig had been in charge of security for the airport at Prague where three flying disc projects were housed. Since this area was the industrial heartland for the SS, Landig would have also been familiar with any aircraft projects under development by the SS. Stevens feels that Landig is a good source of information. Landig goes on to describe some other amazing aspects of alleged Nazi technology under development or operating at the workshops located at Point 103.

Landig says aircraft were guided to Point 103 by a special compass, called a Himmelskompass (heavenly compass). This compass did not work on magnetic lines of force generated by earth's magnetic field as most compasses do. This is because the magnetic north pole is not located at the geographic North Pole, but instead on the Boothia Peninsula near Hudson Bay, many hundreds of miles south of the geographic North Pole. Instead of measuring magnetic lines of force with a needle, the Himmelskompass measured polarized light. Using this device as a guide, an aircraft could be navigated to the proper coordinates without the use of a traditional compass and without interference from a defensive device employed at Point 103, the Magnetofunk. Says Stevens:

> The Magnetofunk was a compass in reverse. It generated a magnetic field and so would confuse any aircraft using a magnetic compass. It would influence the compass just enough to throw off anyone following it as a guide. Therefore, enemy aircraft would be passively diverted around Point 103 without the searching aircraft

even being aware of its influence.

But Landig describes other amazing devices and research going on deep within the bowels of Point 103. He states that a kind of alchemy is at work here. This is not the old alchemy of gold making, conspiracy or black cats, but a new alchemy for modern times. Vril energy was being analyzed. Landig goes on to state that this Vril energy was suitable for powering aircraft. It was the same sort of energy sought by the ancient Indians for their flying craft, the Vimanas. Mercury was being used as a partial means of propulsion. Landig then goes on to make an amazing statement, saying that "our Indian friends" are busy reconstructing ancient secrets under security observation from these ancient sources.[46]

The Mysterious "Blue Island" of the Arctic

Associated with Point 103 is a "Blue Island" that was seen by Canadian air patrols, but was difficult to find when the air patrols returned to the same area for further looks at this Arctic anomaly. Stevens says that routine Canadian air patrols originating out of the Peary Islands began encountering something they could not explain. Says Stevens:

> Several independent reports from around 1950 state more or less the same thing. It seems that these patrol flights (their purpose was not stated) would sometimes encounter what they described as an island, surrounded by an icy ring of high mountains. On this island building structures could just be made out. Further, aircraft of some sort which could not be properly identified were seen. These "notable powered aircraft" could be seen landing.
>
> But there was a big problem. All attempts to approach this island for a better look were met with a strange kind of resistance. Upon drawing closer to the island, the aircraft was surrounded by a "thick blue aether" which was impenetrable to the aircraft's radar.

Stevens points out that these Canadian flights were looking

for UFOs and their base and the Americans were making similar flights out of Alaska. The famous UFO investigator Colonel Wendelle Stevens said numerous times that he was part of these missions in the years right after the end of WWII. With the stories of the Blue Island it would seem that some defensive device was being employed that hid the location of the secret base.

Other evidence of a German base in the Canadian arctic is the movement of the famous German U-boat, U-234, which surrendered near Nova Scotia at the end of the war. The U-234 was a Type Xb minelaying submarine. Type X U-boats were the largest submarines built by the Third Reich, displacing 2,000 tons. In 1944 the U-234 was refitted from a minelayer to a transport ship. It was headed for Japan at the very close of the war, but somehow ended up surrendering to Allied forces between Halifax, Nova Scotia and Portsmouth, New Hampshire, on May 14, nearly ten days after Germany had surrendered. The U-234 departed Kristiansand, Norway for Japan on April 15, 1945 but ended up surrendering a month later near Nova Scotia. What was the U-234 doing in Canadian waters when its mission had supposedly been to take cargo to Japan? This craft should have been in the South Atlantic. Had it come from the mysterious Point 103, the Blue Island in the Canadian arctic? It would seem so.

Landig's secret base at Point 103 was said to be comprised of members of the Thule Society, a society that was sometimes at odds with the traditional power of the Nazi party and the Third Reich. They believed that Hitler and the Third Reich had been compromised by evil forces as early as 1933. Therefore, Point 103 was comprised of mystics and scientists from all over the world. In *Goetzen Gegen Thule,* Landig describes a conference at the base that is attended by the following:

> ...a Tibetan lama, Japanese, Chinese, and American officers, Indians, a Black Ethiopian, Arabs, Persians, a Brazilian officer, a Venezuelan, a Siamese, and a full-blooded Mexican Indian.

Note that Landig claims above that American officers, perhaps Navy officers, were present at this secret base in the Canadian

182

Arctic. Landig says that Point 103 is not an extension of the Third Reich, but a Thulean independent group opposed to fascism, working with esoteric groups from around the world that are in opposition to the Judeo-Christian/Masonic World Order. The focus and symbol of this group is the Black Sun insignia.

Landig describes the large gathering of esoteric groups from around the world that dedicated themselves to the Black Sun cause; they wore either red robes or black robes, black robes being for the higher initiates. He said the scene was obscured by a curtain, but behind the curtain when it opened up was a spacious hall which bore a set of steps leading downward. On both sides of the elongated hall ran some benches on which some of the men of the station sat. The aisles were lower and the processional path of the hall continued through four more levels. Upon this path stood a procession of people that for the most part wore red capes. Says Landig:

> At the head of the procession were foreign guests, whose attire also emphasized the strangeness of this assembly. Over everyone dominated the black, helmet-like headdress of the Tibetan Ta Lama, the Japanese next to him becoming small. While the Tibetan wore the already-known black robe, the Japanese officers had their uniforms on as well, however, also having black robes draped around them.

We have quite a curious group that assembled at this secret Arctic base at the end of WWII. We can see how there might be Japanese officers at the base, but the other people are incongruous, especially the Tibetan Lama. As we have seen, group of Tibetan Lamas was discovered in Berlin at the end of the war and there are rumors of a secret Nazi base in Tibet. Also notable is that the Black Sun insignia is also a yantra—a meditation device to focus the eyes on—and its use is common in Tibet and India.

Greenland and the Lab of Death
The secret base in Greenland with the code name "Beaver Dam" is also associated with the so-called Laboratory of Death.

Stevens said that he first heard about it at a UFO conference, and thought that it was located in Norway somewhere. He later discovered that most of the stories put the secret laboratory in Greenland, and that the rumors were that strange experiments went on there, including the grafting of different animal organs and limbs. Another version, he says, is that this fusion is done on a genetic level, rather than surgical. Obviously the whole idea is horribly creepy and one hopes that there is no such activity.

Stevens says that the fundamental reference concerning the Laboratory of Death is Claude Schweikhart and his 1952 article in the German periodical *Mensch und Schicksal*. He says that the article does not mention genetic manipulation or cloning or chimeras. Schweikhart reminds us of the investigation of the SS undertaken by the Louis Emerich team in conjunction with the Yugoslavian intelligence agency. This is the group that coined the word "Dritte Kraft" (Third Power) for the SS survival groups. Emerich and the Yugoslavians concluded that a base did exist somewhere in Greenland which they called "Laboratoriums des Todes" (Laboratory of Death) but this base remains a complete mystery. Emerich and his group connect the secret base and laboratory with satanic activity.

But the big news from Schweikhart's article is that at the Laboratory of Death in Greenland developed a special type of flying craft in 1951 (others had been developed earlier). This new flying craft had the ability to submerge in the ocean and emerge from the ocean to take flight into the air. In other words, it was a flying submarine. Says Stevens:

> Schweikhart was writing in 1952 so when he says this development took place in an unrepentant Nazi secret base on Greenland in 1951, we can only assume Schweikhart was well connected to say the least. I do not expect any reader to take this statement, a statement from many years ago, from a mysterious source, describing an incredible event, at face value. But I would ask the reader to suspend judgment of the combination flying saucer-submarine until we can deal with that subject in some detail.
>
> So picture, if you will, a base capable of such things.

184

This base would have to be large indeed to accommodate both U-boats and flying discs. Bergmann describes what a Greenland base must have looked like as being essentially the same type of base needed at the South Pole at Neuschwabenland. It would consist of tunnels bored through ice and solid rock to a length of 2,000 meters. The entrance would have been camouflaged so as to make it unfindable.

According to Bergmann, this type of base would have been protected by two kinds of weapons, both called "Strahlenwaffen" or ray weapons. First, there was the weapon he calls "Zuendunterbrechung" or ignition interruption in English. This word is simply a variant of "Motorstoppmittel," called magnetic wave by the Americans. This was a device that ionized the air within its range of operation and so short-circuited the ignition systems of enemy aircraft. The second weapon "Strahlenwaffe" goes undescribed by Bergmann but we may conclude it was a more directed beam weapon since Bergmann goes to pains to describe the result of the magnetic wave-type weapon.

To flesh out our conception of such a base, we must assume parallel tunnels that are interconnected as they were in the underground facilities in Germany, Poland and the Czech Republic. Storerooms, living quarters, power generating facilities, heating facilities, and workshops would also be necessary. In addition, some sort of turntable for incoming U-boats would have been necessary in order to service them and then spin them around for launching again seaward. Access to a rough airfield would also probably be necessary for supplies and reinforcements from Norway.

Stevens thinks that "Beaver Dam" was a submarine base deep inside a hollow cavity in solid rock that was partially above water and partially below water:

In the absolute maximum of possibilities for Beaver

Dam, I would picture it as breakwater or near breakwater, allowing only the tops of the waves into a central harbor. The central harbor would remain relatively calm for a U-boat, especially if it were submerged. To access the base itself the U-boat would dip a few feet underwater under an artificial ledge built into solid rock. Behind the ledge the U-boat would surface into a large hollow cavity completely cut off from the surrounding environment. The cavity would contain lights, dry-docks, machinery for on and off loading. The artificial breakwater and the reinforced cement ledge would be camouflaged to look like real rock or real rock might even be affixed on to it for this natural look. To aerial reconnaissance or even ships sailing past it, the base would blend into the natural surroundings. A small airfield was probably nearby since contact with Norway via air was necessary at least in the beginning.

So, starting in the early 1950s, this facility in Greenland— run by the SS—developed a flying U-boat. These flying disks and cylinders are essentially submarines and airships, both UFOs and USOs. Some researchers even postulate the origin of the UFO that crashed at Roswell as coming from this base in Greenland. This is in agreement with the central thesis put forth by Dr. Joseph P. Farrell in his book *Roswell and the Reich*.[39]

Stevens tells us that other weapons were kept at Beaver Dam according to Dr. Milos Jesensky and Mr. Robert Lesniakiewicz in the 1998 German language book *Wunderland*. Among these weapons was the "Urzel." The Urzel was a combined V-1, V-2, mounted in a horse and rider configuration. This weapon was to be launched at the east coast of the USA and was to have contained an atomic warhead. Unfortunately for the Nazis, the atomic production facilities at Der Riese simply either could not produce the desired weapon in time or were overrun by the Soviets before they could do so.

Stevens says that Dr. Jesensky and Mr. Lesniakiewicz say that the mission and scope of Beaver Dam changed somewhat corresponding to the rapidly changing war situation. They list three stages for Beaver Dam:

In stage one, Beaver Dam would have been the impregnable fortress Dönitz promised for Hitler. Hitler was to be able to carry out war outside of Germany in this vision so U-boats and even tug boats would be necessary since containers carrying large rockets capable of hitting the United States would have been brought in. This base also functioned as a meteorological base until the bitter end of the war in Europe. Training for the men as well as testing and working out of details involving Arctic living would have been carried out at the polar research station at the Golden Peak, Kirkonoshe, in the Bohemian mountains.

In stage two the atomic warheads and very long range rockets could no longer be counted upon since the research, development and industrial facilities lying underground in the Owl Mountains of Poland were unable to provide them in a timely manner. Therefore, Beaver Dam took on a secondary function. It became a repository. Art, artifacts, gold and money from the Reich's Bank, blueprints for superweapons as well as prototype aircraft were brought to Beaver Dam for storage. Among these prototypes/ superweapon blueprints was a flying disc.

In stage three the war is over. The base now served as a hide-away for prominent Nazis on the run but the most important function of Beaver Dam centered around the flying discs. This was their base and point of origin for spying missions directed at stealing atomic weapons and technology from the USA and USSR. Urzel weapons were also in storage. The Nazis at Beaver Dam still prepared for some sort of final battle, a Goetterdammerung, in which they would fight with advanced weapons and machines of all kinds.[46]

Stevens says that these researchers believed that the Germans at Beaver Dam in Greenland wanted to cause provocation between the East and West. Reinhard Gehlen, who ran a German intelligence network inside the Soviet Union during the war and was recruited by the CIA after the war to reconstitute his SS spy network, is often accused of playing one side against the other, even manufacturing the Cold War, for the benefit of those he served in the Nazi period and their ideals. For the SS, the communist Russians were the enemy and the Americans were potential allies.

Stevens says that the Slavic researchers do not think Beaver Dam exists today. He comments:

> At some point during the last 60 plus years they believe the base ceased to function. They really do not say what might have happened but they do speculate about walking through the underground tunnels and finding abandoned, rusty flying discs and derelict U-boats. These researchers believe the evidence is there, only awaiting discovery. Imagine if it were true, if someone did find a secret German base, accessible only by tunnel to the sea, filled with U-boats and supporting machinery that had been in operation after 1945? That would settle the question of postwar German bases once and for all, right?

Stevens says the German researcher O. Bergmann, writing in 1989, says that just as Admiral Byrd led an expedition against the German base in the Antarctic, at Neuschwabenland, so too were expeditions mounted in Greenland, which Bergmann hints were for the same purpose. During the years 1947 to 1950 a Danish expedition commanded by Eigil Knuths was sent to Greenland. In the years 1947 to 1951 the French sent an expedition. In 1951, the English got into the act with their own expedition to Greenland. Says Stevens:

> But what is most amazing is that on March 27, 1951 the Danish, who own Greenland, found it necessary to enter into an agreement with the United States for the common defense of Greenland. Who were they defending Greenland against? Greenland is a sparsely populated land with no large cities, mostly inhabited by Eskimos still practicing variations of a hunter-gatherer lifestyle. The only military power in 1951 set up in opposition to the United States was the Soviet Union and it hardly had its

sights set on invading Greenland.

Immediately, the US government began feverishly building some air bases in Greenland. Bergmann presents a map of Greenland showing three bases, Narsarssuak, Thule and Station North. The base at Station North is located exactly where Dr. Solchec, Dr. Jesensky and Mr. Robert Lesniakiewicz think Beaver Dam was located, on the northeast peninsula called Peary Land.

So what happened here? The Germans were said to have just perfected their flying submarine at this time. How could the Americans and presumably NATO have a base in this exact spot or within a few miles of it?

Stevens says that a likely scenario is that the Germans were displaced sometime in the years 1951 or 1952. They could have abandoned this base or these bases or they could have been forced out by the American military. They might even have entered into an agreement to cede the bases. If this is true, the March 27, 1951 agreement may have involved a secret party, one representing the interests of the Nazis. The hidden base itself with its hidden access to the sea may have been appropriated as a listening post and for use in clandestine activities directed at the Soviet Union.

And we now have an explanation for all of the flying disks seen around the world starting in 1947 or so. Many of them had been made at the facility in Greenland according to Bergmann, and the first of the flying submarines were made as well: cigar-shaped craft that fly in the air or dive into the water. Another one these factories probably existed in Antarctica as well as one in Argentina.

A Strange Encounter in 1947

Coral Lorenzen was a UFO researcher and author who founded NAPRO (National Aerial Phenomena Research Organization in 1952. She attempted to keep an open mind about UFOs during her decades of research and said in the preface of her 1962 book: "An underlying premise with me is that the UFO problem is planetary in scope—it knows no national boundaries. It is my feeling the evidence should be allowed to provide the answers; therefore I present the evidence with as little prejudgment as possible and suggest such meanings and patterns as seem to be valid." In 1962 she published her important UFO book, *Flying Saucers: The Startling Evidence of the Invasion From Outer Space.*[17] This book chronicles many important, and often bizarre, UFO incidents in the 1940s and onward.

Coral saw her first UFO in the small town of Barron, Wisconsin in 1934 when she was nine years old. The sighting was during the day and she said that the silvery disk moved in an undulating fashion to the west-southwest in a clear sky. This sighting has never been explained. Later, on June 10, 1947, she was married and living with her husband, Jim Lorenzen—a pilot in the Air Force—in Bisbee, Arizona. On that night she put her young daughter to bed and then watched as an orange glow could be seen to the south of town, along the Arizona-Mexico border. She says that she watched as a glowing orb rose from the ground in the distance and rose straight up into the sky and disappeared. This was her second UFO sighting and she was intrigued by the mystery of what she had seen.

And then on June 24, 1947, the businessman and private pilot Kenneth Arnold saw nine objects moving in a skipping motion over the Cascade Mountains of Washington State. Arnold's sighting became famous and Lorenzen began her collection of data on UFOs which was to ultimately become part of APRO and her books. On June 27 more UFOs were seen in the area of Bisbee and she cataloged these sightings as well. Lorenzen's book begins with these early sightings and goes from there.

Throughout her book *Flying Saucers* she essentially recounts the various UFO sightings she has collected in chronological order starting in 1947 and continuing to the early part of 1966 (in the second edition). She puts particular emphasis on the years 1952, 1954, and 1965 as there were a number of UFO "flaps" during these years of heavy UFO activity.

The first story she relates is a curious one that seems to fit into the now-known lore concerning the Haunebu, Vril and Andromeda craft allegedly designed and created during WWII in Germany. Although the event occurred in June of 1947 she did not learn of it until 1954 when she received a letter from a Mrs. A.M. King of

South Africa. The letter to Lorenzen said:

> I left Mombasa [Kenya] at the end of June 1947, on the SS *Llandovery Castle* en route to Cape Town and, going through the straits of Madagascar about the beginning of July, was on deck with another lady passenger at approximately 11 PM when we noticed a particularly bright star. It was traveling very fast and approached the ship. Suddenly a searchlight appeared which flashed a strong beam of light on the water within fifty yards of the ship. It descended, its beam shortening and becoming brighter as it neared the water, and the next instant there was no more light, but an object appeared, apparently made of steel and shaped like a cigar cut at the rear end. It remained in the air about twenty feet above the sea, parallel with the *Llandovery Castle*, and traveling in the same direction.
>
> Gaining a little in speed, after a second or two the whole shape disappeared without a sound, from the rear issuing fierce flames which shot out to about half the length of the object. It appeared that there must be something like a huge furnace inside the thing, but still we could hear no noise from the flames. No windows could be seen, only a band of metal around the entire thing which, if it had been a complete cigar shape, would have been centrally situated.
>
> The object was very large, about four times the length of the *Llandovery Castle* and at a rough guess about four times as high. We had a wonderful view, but in a few seconds it disappeared. No light was seen forward as it left; it just vanished soundlessly in the darkness. For a while we thought we were the only ones on deck at that late hour but, walking to the prow of the ship, we saw there one of the ship's officers with a few passengers; the entire party had seen the same thing. Whether or not it is recorded in the ship's log, I know not.[17]

Coral Lorenzen then comments: "If Mrs. Kings estimate of the object's size in comparison to that of the ship is fairly accurate, we must assume the object was at least sixteen hundred feet long and possibly larger (ocean-going vessels vary considerably in size)."

She then discusses the January 7, 1948 case of National Guard pilot Captain Thomas Mantell who took off at 2:45 AM from Godman Field in Kentucky to chase a UFO seen over the area. Mantell approached the object, which he said was enormous in size and apparently metallic. He then apparently passed out from the altitude and crashed his plane around 3:50 AM and was killed. He is thought to be the first pilot to chase a UFO and then be killed in a crash. The Air Force tried to explain the incident as an unfortunate case of a pilot chasing the planet Venus, thinking that it was a UFO. Lorenzen did not buy this explanation.

The craft described by Mrs. King is identical to the cigar-shaped submarine craft supposedly being built in Greenland as well as the cigar-shaped craft photographed by George Adamski. This was in 1947, basically two years after the end of the war, and Mrs. King was apparently witnessing one of the Andromeda craft built in Germany at the end of the war, or another craft built by the Germans in Greenland. Was the craft on its way to Antarctica after flying around parts of East Africa or the Middle East? This intriguing sighting is probably one of many in 1947, with only a few making it into books or newspaper articles. Soon such witnesses would be labeled as quacks and hoaxers, and this would discourage many people from telling their stories. Fortunately, many brave souls came forward with their tales and we are fortunate to have Coral Lorenzen to help bring such witness statements to the public's attention.

> Argentina remains a mysterious maze
> of questionable characters.
> —FBI Document, September 4, 1944

Strange Things in Argentina

Whatever craft was being built in Greenland seems to have made its way to other secret cities of the SS in other parts of the world. One may have visited the SS airbase in Tibet, but it seems that most of these craft were sent to the other pole, to Antarctica and Argentina. Haunebu and submarine activity must have been occurring during 1947 and 1948 but there are very few public reports. Much of the Nazi UFO activity at this time was probably focused on remote areas of Patagonia from whence little information was likely to leak from at any time. The remnant of the Reich couldn't just exist in the icy and remote bases in the Arctic and Antarctic. It needed real territory where there were farms, ranches and factories. For postwar Nazis, this was Patagonia, other areas of Argentina, Paraguay and Chile. Uruguay was never really a Nazi investment area, but there are large German populations in southern Brazil.

The first Haunebu or UFO report in Argentina was in 1949, four years after the end of the war. This incident occured in a town called El Maitén in Chubut province in northern Patagonia. In 1865, Welsh people came to Chubut and settled in the Chubut Valley area, which became one of the most prosperous provinces in Argentina. Volga Germans followed them into the area and the large Volga German settlements are in valleys just to the north of Chubut Valley. All of these ethnic communities learned to speak Spanish but kept their ethnic identities and languages. The Welsh of the Chubut area would not have seen themselves as allies of Britain, but simply as Argentinians.

El Maitén started as a rural community but was influenced greatly by the arrival of the General Roca railroad in the area in 1939, on a branch that continued to Esquel. This branch was completed and opened in 1945, and El Maitén was selected as the site of its maintenance sheds and locomotive warehouse. This was one of the southernmost stations for the extensive railway network in Argentina, nearly all of it built by the British.

According to the account on Wikipedia, on February 20, 1949, witnesses in the town said that a flying saucer descended from the sky and landed in the town itself. After the flying saucer landed, three normal-looking men walked out of the craft. After a short time they returned to the craft and it took off, ascending up into the sky and out of sight.

The town at the time barely had a police brigade, a guard station, a railroad shed beside the narrow platform and a half-finished house. There were also some small farms in the vicinity. It would seem that this small town along the Andes near the border with Chile was a safe place to land a craft and have a look around. Perhaps someone was waiting for the craft in this remote locale, or the occupants had simply stopped their craft here for other reasons. Was this a Haunebu or Vril craft coming from a base further south or even from Antarctica? This seems likely.

Argentina's next big UFO encounter that was promoted in the media was the sighting of a flying saucer in April of 1950 in Resistencia, a city in the northeast, very near to Paraguay. The newspaper *El Nacional* of Resistencia reported on April 18, 1950:

> Around noon today, the people of Resistencia, without exception, looked to the sky to see a "flying saucer" that according to some, vanished toward the southwest in the horizon, after tracing a curve in the sky. According to others, however, it was nothing more than Venus, which was a great cause for conversation in Paraguay a few days ago. The fact is that the widest variety of conjectures and theories was put forth about this subject. The comments were the talk of the day in every meeting.
>
> Not a few people awaited the landing of the famous disk after what was reported by *El Nacional* regarding the event that occurred in Patagonia. But the "disk" went away, leaving doubts in its wake and a knowing smile on the mouths of incipient local astronomers.

One has to surmise that this was probably a Haunebu. But things were to get even more exciting in 1954 as Coral Lorenzen chronicles in her book.

Strange Things in South America—1954

In her 1961 book, *Flying Saucers,*[17] a chronicle of her collected UFO incidents from APRO members, Coral Lorenzen has a chapter: "1954—Europe and South America." This has some important UFO incidents that apparently involve the Haunebu, especially in South America. Says Lorenzen of the many incidents beginning in October of 1954:

> On October 23 a disk-shaped object was seen by

workers and policemen as it maneuvered over Buenos Aires, Argentina, stopping and changing course many times. The object gave off luminous rays which changed colors alternately. On October 24 air-base personnel in Porto Alegre, Brazil [in the furthest south of Brazil, near the Uruguay border], observed objects which they could not definitely identify as conventional phenomena—they were circular, silvery in color and maneuvered at great speed with occasional abrupt changes in course. In the Chicama Valley of Peru an engineer observed a brilliant elliptical object on October 27. It pulsated brilliant flashes of light and for several minutes it moved slowly, at other times with great speed and at one time it fell diagonally. The engineer reported that the object hovered at a height of three hundred meters and recovered its luminosity which had paled during the diagonal fall. The orchestra was tuning up.

And then on November 10 a Porto Alegre agronomist out for a ride with his family saw a disc from which emerged two apparently normal-shaped men with long hair and over-all-like clothing. They approached the car with their arms above their heads, but the driver, urged on by his wife and daughter, accelerated and left the strange men behind. The motorists saw the men enter their disc-shaped craft and mount the sky at a dizzying speed. In Curitiba, Brazil [just north of Porto Alegre], a railroad worker told authorities that at 3:30 a.m. on November 14 he saw three beings in tight-fitting, luminous clothing examine the ground around the railway tracks with a lantern. When the strange creatures saw the man they entered an oval-shaped craft which elevated rapidly. On the same day a bright oval object giving off a yellowish glare was seen by many in Buenos Aires as it traversed the sky. Also on the 14th, three hundred members of an anti-aircraft division in Berna, Argentina, watched a silvery, disc-shaped object give off a reddish trail. Observers using binoculars said it appeared to be at nine thousand feet when it was first spotted hovering in the air; suddenly it began to fly south at great speed, disappearing into the clouds.[17]

According to Lorenzen, the craft are often seen flying south, probably headed for a German base in southern Argentina or Antarctica. These sightings so far fit very well into the German flying disc scenario, including the description of two normal-looking men in overalls walking toward a car with their hands raised above their heads. Were they possibly Germans, and not just Germans but veterans of World War II? Were the overalls the standard dress at the Antarctic base(s)? One can imagine that these post WWII bases in remote parts of the world did not have barber facilities, so long hair (blonde?) would have been normal for these pilots, as well as in the U-boats. Shaving would have been an easier daily ritual and it seems that all the UFOnauts, tall and short, had clean-shaven faces.

It is also curious to note the Peruvian encounter above in which the disc "hovered at a height of three hundred meters and recovered its luminosity which had paled during the diagonal fall." This would seem to be the rebuilding of the electro-gravitation field around the saucer as it hovered after its diagonal dive, at which time the craft took off again at great speed. And why would a flying saucer go through such aerial antics if not for the reason of showing off its aerial capabilities to an observer? It seems that in most of these cases the craft know they are being observed and sometimes photographed and they want this. They want people to see them and they want to be in the news. In South America, at any rate, they did get into the news.

Lorenzen then finishes this chapter about 1954 with some curious stories from November and December that bolster her belief that these craft are extraterrestrial and from Mars, often taking samples of plants and the soil starting with a curious encounter in Venezuela that reinforces the claim that the Nazis and the SS had a radio station and Haunebu base somewhere in that region. Says Lornezen about the incident in December 1954:

> In the early part of December (unfortunately there is no exact date for this incident) a respected and well-liked teacher and director of an educational institution

in Barquisimeto, Venezuela, related how he was pursued by a glowing, disc-shaped object while driving along the highway to Guanare. The object when first seen resembled the moon, and he gave it no further thought. He then noticed that the object was moving toward him and was huge in size. Seized with panic, the professor fired a small revolver he carried in his car at the pursuing object as it maneuvered above his car. The bullets didn't seem to bother the object and, thoroughly frightened, the man pushed the accelerator to the floor. He managed to stop another car occupied by a lawyer, a sheriff and a policeman, bound in the opposite direction. They accompanied the professor back to where he had last seen the object, and they were just in time to see the object heading into the southern sky leaving a bluish trail behind.

On the evening of December 10 a well-known Caracas doctor witnessed a strange sight while driving with his father in the vicinity of Floresta. He was later interviewed by the press, though his identity was withheld since, like most observers, he did not care to be ridiculed for what he saw. At 6:30 p.m., between La Carlota Airdrome and Francisco de Miranda Avenue, his father suddenly pointed. The doctor stopped the car. Together they watched two little men running from the brush. Shortly after they disappeared in the thickets a luminous, disc-shaped object emerged from behind the brush and, with a sharp "sizzling" sound, darted off into the sky at a high rate of speed.

On that same day an American engineer employed by a petroleum company in El Tigre took snapshots of a covey of five saucers as they flew in formation from south to north. They resembled "turtles," with a beam of light at the front and another at the rear of the formation. The photos were shown to the police and the press and were discussed by various newspaper articles; though they were not published. The description of these objects tallies with the objects photographed by a Salem, Massachusetts, coastguardsman in 1952 and the famous Trinidade photo of January, 1958.

A group of prominent professional and scientific men in Venezuela realized by now that something was definitely afoot and they set about attempting to find an answer. Horacio Gonzales was one of them. Another, Astronomer Dr. Aniceto Lugo who publicly professed his interest in UFO, had more than a passing interest in them, for in September, while Europe was under close scrutiny of hovering disc-shaped objects and little men, he had observed flashes of light inside Crater Kepler on the moon. A similar observation of unidentified light flashes of one to two minutes' duration in Crater Copernicus was made by Mr. Venegas, secretary to the Minister of Communications.

The peak of the sightings, whether of objects or little men, appears to have been reached during December 10-14. Brazil also was being initiated, the Brazilian Air Force making public an appeal for the united efforts of all governments toward the solution of the UFO problem. Caracas, Venezuela newspapers devoted whole pages to a briefing of Brazilian Air Force officials by Colonel Joao Adil Oliveiera who, incidentally, gave good coverage to many "classics" in UFO history. The Brazilians had good reason to be excited.

On November 4 Jose Alves of Pontal was fishing in the Pardo River near Pontal. The area was deserted, the night quiet with only a slight breeze blowing from the east. Suddenly, Alves spotted a strange craft in the sky, apparently heading toward him. He watched, transfixed, as it closed in with a wobbling motion and landed. It was so near he could have touched it, he said. The object appearing as two washbowls placed together, looked to be about ten to fifteen feet in diameter. He was too frightened to run. Three little men, clad in white clothing with close-fitting skull caps, emerged from a windowlike opening in the side of the small craft. Their skin appeared to be quite dark. Alves stood terror-stricken, watching the small creatures collect samples of grass, herbs and leaves of trees; one of them filled a shiny metal tube with river water. Then, as suddenly as they had come, they jumped back into their machine, which took off vertically as swiftly and as silently as it had come. Residents of Pontal, who heard Alves's story when he came back to town, told the press that he was a quiet man who lived only for his work and his family. He had never heard of flying saucers and he was sure the little men were some kind of devils.

The Air Force base at Santa Maria, Rio Grande do Sul, was the scene of a strange incident on November 22. At 9:45 p.m. radio operator Arquidmedes Fernandez left the meteorological station and walked toward a small building a hundred feet away that he had built to store the thermometers and other instruments. Gathering readings for the next weather bulletin, he took a customary look at the sky and noticed a thin dark cloud hovering above small eucalyptus trees behind the building. He observed it very closely. It was not a cloud—it was a black object, enormous in size, shaped like a washbowl hanging upside down, and seemed to be about 160 feet long, suspended just above the trees. Not motionless as a cloud would appear to be, it had a slow, oscillatory motion. Fernandez became alarmed and ran back to the station, watching the object all the time. It slowly lowered itself into the trees and then began climbing again, rather swiftly. He then noticed a small light on top of it. It dived again between the trees, almost to the same place he had seen it originally. The object moved again, now glowing faintly in the darkness. Fernandez radioed Porto Alegre and reported what he was seeing; the object was still there. At 1:15 a.m. his substitute arrived. Fernandez had watched the object for almost three and a half hours. Fernandez's startling report was corroborated by the testimony of others who also saw the strange object. At midnight another radio operator, Ruben Machado, had seen it from the window of his room in the Canobi Hotel, some distance from the base. To him it appeared to be hovering over the base—a luminous object lager than the full moon. He pointed it out to others who saw it clearly before it disappeared into the north. There was further substantiation when Varig radio operator Jurandir Ferreira reported he had spotted the UFO over the base. He saw it as it was gaining altitude; after some maneuvers it headed into the north and was gone.

On the evening of December 9 a farmer, Olmiro da Costa e Rosa, was cultivating his French bean and maize field in Linha Bela Vista, two and a half miles from Venancio Aires, Rio Grande do Sul, when he heard something with the sound of a sewing machine. The animals in the pasture next to the field scattered and ran. Costa e Rosa looked up and saw a strange-appearing man. Beyond him was an unusual object hovering just above the ground. It had the shape of an explorer's hat, cream-colored and surrounded by a smoky haze. There were two other men, one in the craft, his head and shoulders sticking out, the other examining a barbed-wire fence. Costa e Rosa dropped his hoe and the stranger nearest to him raised his hand, smiled and picked up the hoe, which he turned in his hands, examining it carefully. Then the man placed the hoe in Costa e Rosa's hand, bent down, and uprooted a few plants and started toward the craft. Costa e Rosa had stood as though paralyzed. Then, assuring himself that they meant him no harm, he advanced toward the craft. The man who had taken the hoe and the one in the craft made no move to stop him, but the one at the fence made a gesture which seemed to mean that he should stop. Costa e Rosa stopped. Some of the farmer's animals approached and the strangers looked at them with great interest. With words and gestures the farmer tried to tell them he would be happy to make a gift of one of the animals. The strangers didn't seem enthused about the offer. The departure of the strangers was as unexpected as their appearance. Suddenly they trooped into the ship which rose above thirty feet, accelerated abruptly and flashed away into the western sky at high speed.

Costa e Rosa's description of the men was detailed—he had had time to observe them at fairly close range. They appeared to be of medium height, broad-shouldered, with long blond hair which blew in the wind. With their extremely pale skin and slanted eyes they were not normal looking by earth standards. Their clothing consisted of light brown coverall-like garments fastened to their shoes.

Afterward Costa e Rosa said the shoes seemed especially strange because they had no heels. After the men had left he searched the ground over which the objects had hovered but found nothing. However he did notice the smell of burning coal which remained in the air for some time after the craft's departure. Costa e Rosa was questioned for several hours by authorities from Porto Alegre. It was determined that he didn't read science fiction; indeed, he read with difficulty and had never heard of "flying saucers." He seemed to believe that these men were visitors from another country. This incident was reported in the magazine *O Cruzeiro* by one of Brazil's outstanding reporter-writers, Joao Martins who is also a crack UFO investigator.

Two days later at 5 p.m. Pedro Morais, who lived less than a mile from Costa e Rosa's home, was preparing to go to a warehouse for supplies. He heard the frightened squawks of a chicken, and he went out to investigate. The day was hot, with no wind. He still heard the chicken but couldn't find it (and never did), for what he saw hovering in the air took his mind off the chicken and the sparrow hawks he had thought were molesting it. The object had the sound of a sewing machine; it oscillated as it hovered and appeared on the topside to be shaped similar to the hood of a jeep. The bottom resembled an enormous polished brass kettle. Morais's attention turned to the cultivated fields nearby where he saw two human-shaped figures. Indignant at this trespass, he started toward the craft. One of the men started running toward him, while the other raised his arm in a gesture which appeared to be a warning not to come any closer. Morais, still angry, did not obey, and continued toward the machine. He noted that as one ran toward him, the other kneeled down and quickly picked a tobacco plant out of the ground. Then both got into the craft which disappeared from view in the sky within a few seconds. Morais said the little men were human in shape but didn't have any faces. He got the impression that they were enveloped in a kind of yellow-colored sack from head to toe. After their departure he looked for footprints, but found none. But the hole from which the plant had been uprooted was still there. Morais, too, turned out to be a rather simple, uneducated man; he did not know even the alphabet. But, unlike Costa e Rosa, he thought the men were saints or ghosts. When he was told that the government was anxious to have one of these "men" dead or alive he vowed to shoot one if he ever had another opportunity.

And so the picture of the 1954 saucer invasion of South America is fairly clear—first, a scattered observation, then the landings; the pattern is very similar to that of the European visits in September and October. The incidents reported here are only a sampling of the hundreds brought to public attention by the press in Brazil and Venezuela as well as in other South American countries. The seeming lulls may be due to our inadequate coverage of other areas on the continent; however, the many reports which did come to our attention were startling in that most of the sightings were in the vicinity of defense installations. The possibility that these objects, which had kept their distance in years past, were merely American or Russian secret weapons became invalidated. No reasonable, intelligent individual with any command of the facts would any longer fall for that one.

And so Coral Lorenzen concludes the 1954 invasion, which she believes involves extraterrestrials from Mars. Neither she, nor her husband in the US Air Force, have any knowledge of secret German projects, advanced technology or the activities of the SS after the war. They are simply trying to make sense of the reports that they receive.

More Argentina UFO Encounters
One of Argentina's best-known UFO cases occurred on July 3,

1960, and the witness was a high-ranking officer of the Argentinean Air Force (AAF), who by a fluke was able to obtain a remarkable photograph as supporting evidence of his encounter. The case was not given much publicity until 1977, when it finally was reported in South American UFO magazines, which are popular on that continent.

The witness in this sighting is Hugo F. Niotti, then a captain of the AAF assigned to the Air Force School for Sub-officers located in the city of Cordoba. Contrary to what many would expect, his involvement in the case did not affect his military career, and seventeen years later, when finally interviewed by a UFO magazine in Argentina, he had risen to the high rank of vice-commodore, occupying a responsible position within the AAF.

On July 3, 1960, then Captain Niotti was driving in central Argentina from the town of Yacanto toward Cordoba, Argentina's second largest city. After Buenos Aires, Cordoba is the most important city in Argentina, with a large population and an industrial base. In 1954 the city became the site for Argentina's automakers, which included German companies like Volkswagen and Mercedes. It is likely that Cordoba, with automakers and other industries creating and importing all sorts of mechanical parts, was a source of the material needed by the secret German factories. Some of these factories could have been located in Patagonia or Chile, but some could have been in the busy city of Cordoba

The photograph taken by Captain Niotti.

A close-up of the photograph taken by Captain Niotti on July 3, 1960.

itself. Indeed, this north-central city was well connected to other South American cities and would be a good place to have a covert business or two.

On the day of Captain Niotti's encounter, there was a slow drizzle and the road was rather slippery. At about 4:30 in the afternoon Niotti had just finished negotiating a wide S-curve, when he suddenly noticed a rather close and unusual object hovering near the ground to the right of the road.

Startled, he stopped the car, grabbed his camera (fortunately next to him on the seat), got out and moved a few steps away, and proceeded to take a photo of the object, which was moving slowly. While he was engaged in winding the film to take a second shot, the object started to accelerate and disappeared into the clouds, which were very low. Captain Niotti jumped back into his car and continued his trip to Cordoba, where he proceeded to have the film processed.

Niotti said the object was conical in shape, with a height of seven to eight meters and a base diameter of three to four meters, with its axis almost parallel to the ground and its base facing the witness. It was at a distance of 80 to 100 meters from his location and moving very slowly toward the south, always parallel to the ground. The craft rotated very slowly. It then accelerated very rapidly, and disappeared into the low cloudbank. This sudden acceleration without any sound was inexplicable to the witness in view of his proximity. The color of the object was a uniform dark gray. He said the surface was perfectly smooth without joints or rivets and had a definite metallic aspect. A horse can be seen in the photo, turning to look at the object.

The city of Cordoba is near the famous UFO town of Capilla del Monte, and as we have said before, Cordoba is the focus of the German-Argentinian "military industrial complex."

Men in Black in South America

A curious story is told by Jim Keith in his book *Casebook on the Men in Black*,[31] about a Spaniard who was visiting Venezuela and had a strange encounter. Keith says that Guillermo Arguello de la Motta, a Spanish physician, was visiting friends at a small town near Caracas, Venezuela in July of 1971 when he observed two men in black suits, red ties and black berets get out of a red Mustang sedan and then stand there by the car. Says Arguello in Keith's book:

> The two men stood there waiting for about five minutes and then began to put on orange-colored belts, talking together animatedly in the meantime. Suddenly a shiny object appeared in the sky. It rapidly descended and then stopped at a height of about 60 centimeters from

A Vril craft photographed over Milan, Italy on February 12, 1960.

205

the ground. It was circular, bell-shaped underneath, and with a turret on the upper part. Its width could have been about 30 meters. What surprised us most of all about it [were] the rapid changes of color, from orange to blue and then to white. When it halted, floating in the air, it rotated through almost 180 degrees. Suddenly a small parabolic staircase came down from the base of it, which enabled the two men from the Mustang car to enter the saucer with ease. When the staircase had been drawn in again, the craft dipped slightly towards its left side and then, following an inclined flight path, vanished into the sky at an impressive speed. The machine was of course definitely no helicopter. It was totally silent, and its shape was something totally unknown, i.e., not conventional.[31]

The witness never did see who was inside the craft, but the two men in black suits and black berets—Men in Black—were simply waiting for the craft as if it were a city bus, put on orange belts and hopped on board. Ride in a flying saucer, anyone? Stevens says that during the 1950s, 60s and 70s it was necessary for secret SS officers and Project Paperclip Nazis to travel in this manner, rather than risk travelling through major airports with forged passports. Many of the former SS officers who had served in the war were getting quite elderly by the 1970s and ultimately these last officers came in from the cold, so to speak.

Jim Keith describes another curious encounter in Latin America, this one near Mexico City on May 3, 1975. Keith says that a private aircraft pilot named Carlos de los Santos nearly collided with three disk-shaped craft while traveling in a small plane near Mexico's large capital city, an encounter that was confirmed by radar operators at the airport. The next week de los Santos claimed that he was driving to a television station to discuss his close encounter with the three disks, an appointment he missed, when he was forced to stop on the highway by two black Ford Galaxie limousines. Says Keith:

> Four men in black suits approached de los Santos and warned him in Spanish in a "mechanical tone" to be quiet

about the incident "if you value your life and your family's too." A month later on his way to visit with American scientist and ufologist J. Allen Hynek, de los Santos met an MIB on the hotel steps. The MIB shoved him and made a statement leading de los Santos to believe that he had been under observation. Again, de los Santos missed his appointment.[31]

In that same year, 1975, an Argentine anthropologist and UFO investigator named Anton Ponce de Leon described the strange occurrences that happened to him while investigating some UFOs around a small Peruvian town named Sicuani. Sicuani is a town on the main highway—a two-lane paved road—between Cuzco and Lake Titicaca. It is a major truck stop town with numerous gas stations, restaurants, hotels and shops—but still not a particularly large town. Ponce de Leon describes his investigation thusly:

> In the year 1975 there was much talk about UFO sightings in Sicuani, particularly at night, and the peasants were talking about something very worrisome to them, that they were hearing noises under the ground as if a machine

A photograph of three disk craft over Sicuani, Peru taken on December 6, 1968.

207

were working there, according to their expression.[31]

Keith says that Ponce de Leon met with a newspaper reporter from Lima, Peru, who worked for the newspaper there called *Ultima Hora*. This reporter told Ponce de Leon that he had film of some flying saucers taken about a week before in Capilla del Monte, Argentina, a northern Argentina town many hundreds of miles to the south. As we have discussed, this area is famous for UFO disc sightings.

The unnamed reporter from the *Ultima Hora* newspaper had photos of the disk craft in Argentina and was now at the Peruvian town of Sicuani with his undeveloped film investigating other UFO sightings when he met Anton Ponce de Leon.

The photographer told Ponce de Leon that two men in black suits had come to his hotel when he was not there and told the clerk that they were his friends and asked to wait in the room for the photographer's return. Instead they ransacked his room, apparently looking for the film, and then left. The photographer had actually given the film to a friend in town to develop.

Later he met the men in black suits at the hotel lobby where they demanded his film. He told them the film had already been picked up and they left, but he found that his room continued to be

A cigar-shaped craft photographed over Puerto Maldonado in the Peruvian Amazon region near Brazil on July 19, 1952 by Domingo Troncoso.

entered and papers that were on a table were taken. He eventually fled the town in a taxi to Cuzco and then flew to Lima where the photos were ultimately published in an issue of the popular Lima newspaper *Ultima Hora.*[31]

This is just a sample of the strange happenings in South America that went on for decades, with numerous UFO sightings and even the strange story of noises coming from under ground like there was machinery at work. Was it possibly a tunnel-boring machine beneath the city? If so, was this some postwar operation being conducted, not by the government of Peru, but of some Third Power of ex-Nazis operating out of Antarctica and Argentina? Is there a secret UFO base somewhere near the northern Argentina town of Capilla del Monte? It would seem so.

The Winged Serpent Insignia

A curious UFO "abduction" incident took place in 1967 involving occupants of a flying saucer wearing caps with an antenna on one ear and the emblem of a winged serpent on the left breast. Was this a new emblem being used by the SS in South America? The winged serpent is typically identified with Quetzalcoatl, the winged serpent god of Central America. The Egyptians used the emblem of the winged serpent as well.

The story begins in Ashland, Nebraska, on December 3, 1967, where Herbert Schirmer, a 22-year old patrolman, was making his usual rounds late at night. He had checked the Ashland Sales Barn, and several gas stations along Highway 6 when he noticed what he thought were red lights on top of a large truck. He had just passed through the intersection of Highways 6 and 63 at about 2:00 AM. He drove the short distance down Highway 63 and stopped with his headlights shining on the object with the lights.

According to Schirmer, the object was a metal oval or disk shape. The red lights that he had seen were blinking through the oval portholes of a metallic, oval-shaped object that was hovering at a height of about eight feet above the road's surface. The object appeared to have a polished, aluminum surface and had a sort of walk around it. It had a structure underneath, like landing gear.

The next thing that Schirmer could remember was that the object rose into the air with electric flames coming from underneath. It

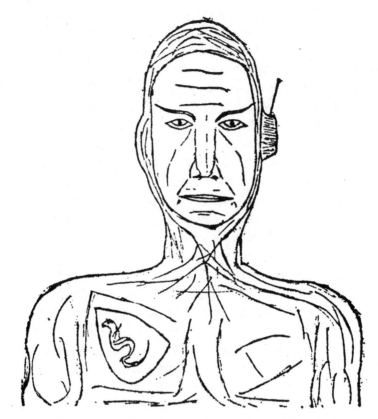

The UFO occupant drawn by Patrolman Herbert Schirmer in 1967.

passed almost directly over Schirmer's patrol car, and the object quickly shot out of sight. Schirmer returned to the police station, noting that it was now 3:00 AM. This surprised him because he felt that only ten minutes had passed and now it seemed that he had about 45 minutes of missing time.

At the station, he made this entry into his logbook: "Saw a flying saucer at the junction of highways 6 and 63. Believe it or not!" Afterward, Schirmer developed a red welt on his neck, a headache, and he began to feel ill. Word of Schirmer's sighting was related to the Condon Commission at the University of Colorado, which investigated the incident.

Schirmer was asked to come to Boulder, Colorado. At Boulder, on February 13, 1968, he was hypnotized by psychologist and UFO researcher Dr. Leo Sprinkle of the University of Wyoming. Under hypnosis, Schirmer recalled that, after he stopped his car near the object, the engine died and his radio went silent. A white object emerged from the craft and seemed to communicate mentally with him, preventing him from drawing his gun as he was wont to do.

After the hypnotic session had ended, Schirmer was able to recall even more details about the encounter. The beings were friendly, they drew energy from electrical power lines, and they told him that they had a base on Venus. He said that the occupants wore uniforms with an emblem of a winged serpent on the right breast as well as a headphone with an antenna coming out of it over one ear. He drew a picture of one of the occupants, who appeared

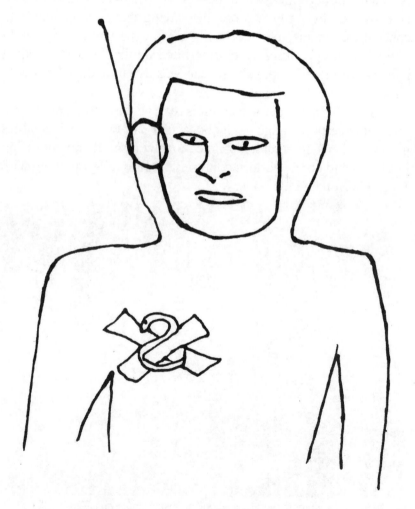

The UFO occupant drawn by Iris Cardenas in 1979.

211

to be an ordinary male human with a special flight suit and head covering with the earphone and antenna over the left ear and the flying serpent emblem inside a triangle over the right breast.

The Condon Committee concluded: "Evaluation of psychological assessment tests, the lack of any evidence, and interviews with the patrolman, left project staff with no confidence that the trooper's reported UFO experience was physically real." Psychologist Dr. Sprinkle, however, felt that Schirmer "believed in the reality of the events he described."

Schirmer returned to his job in Ashland and was appointed Ashland's Police Chief when the former chief resigned. However, he resigned after only two months, saying that he was unable to concentrate on his job due to his UFO experience. Apparently things did not go well for Schirmer; he said that he was ridiculed by some of the townspeople, his car was dynamited, and his wife left him.

Assuming Schirmer's account is real, we have to wonder what this flying serpent badge is all about. Is it part of some special SS group, perhaps one that is designed to sow discord and confusion in the US and other Western countries? Part of a Psyop along with Joseph Farrell and his books?

A Vril craft photographed by a Mexican businessman on a Tucson, AZ to Mexico City flight and published in the newspaper *El Diario* in August of 1965.

Landig tells us that the SS had split from the Nazi party and, in control of all of the secret facilities around the war—the secret cities of the SS—the SS continued their operations out of Antarctica and Argentina. It seems one of their operations was a Psyop against the US, Europe and the West including the secret promotion of a flying serpent brotherhood with its own symbol on a uniform, very much as was done in WWII. The wearing of medals of honor and other badges was to largely be discontinued after WWII and today high ranking military personnel in a battlefield area rarely wear the badge of their status in an obvious spot, but it is there nonetheless.

As to the curious winged serpent badges as described in Schirmer's hypnotic sessions, there is another case where this insignia was used. In the 1982 book *UFO Contact From Undersea*[3] by Lt. Col. Wendelle Stevens (retired) the curious case of Iris Cardenas is explored. Stevens was drawn to the case by a newsclipping from the *Miami News* February 23, 1979 edition. Stevens investigated the case which was already being investigated by a Cuban refugee UFO researcher named Doctor Virgilio Sanchez-Ocejo. Much of the book recounts Sanchez-Ocejo's research which includes multiple sightings and abductions of the Cardenas family, Filiberto and Iris, who lived in Hialeah, Florida.

In their first encounter their car malfunctioned and the engine died. After getting out of the car, Filiberto said he was struck by a beam of light from a flying saucer and paralyzed. His wife, Iris, drew a picture of one of the occupants of the flying disk and it was very similar to the one described by Schirmer: a human with a suit and hood and an earphone over one ear with an antenna coming out of it and a serpent with a large X behind it. The beings' message to the Cardenas was one of Universal Love. They also told her that the Catholic Pope was working with them and was also a man of great love.

So, what are we to make of the saucer men with their antenna ears and serpent emblems? The earphone with the antenna seems very 1960s and 70s. The serpent emblem would seem to indicate that these occupants are from planet Earth and were using the earthly symbol of a snake as the insignia of their "strike group" or squadron. With their paralyzing rays and their uniforms, they

213

appear to be some faction of the breakaway SS groups in South America. Was their mission to create confusion in the United States in regard to the many stories of UFO encounters?

In his book *Saucers, Swastikas and Psyops*,[38] Joseph Farrell discusses the Nazi International and its psyops against the West. Fundamentally important to these psyops was money, and Farrell says that the central figure for this money and the various operations of Nazi International was Otto Skorzeny. Says Farrell:

> It was Skorzeny who played a central role in Bormann's plan to evacuate liquid and hard assets in the form of cash and various bullions out of Europe to safe havens in South America. As such, Skorzeny was an early target of America's war-end intelligence efforts, not so much to bring him to trial for war crimes, but to employ him in its own postwar anti-Communist efforts. As such, Skorzeny is the link between Bormann's Nazi International in South America, Gehlen's postwar group of Nazi spies in Eastern Europe based out of Pullach, and American intelligence, and had in fact even met with America's OSS chief, General William Donovan after the war in Nuremberg.
>
> Yet, as all of these links were forged, Skorzeny maintained contact with the SS elements behind the Iron Curtain that had ostensibly gone over to serve the Communist cause! Skorzeny was also related to Hjalmar Schacht, having married his niece and was in frequent contact with Schacht himself both from his headquarters in Spain, and later during the aftermath of the overthrow of King Farouk. It was during that aftermath, when the Nazis had not yet been thrown out of Egypt by Nasser, that Skorzeny and his Nazis, with Nasser's approval, conducted pogroms and murders against Egyptian Jews, confiscating over $100 million of Jewish property in the "New Egypt."
>
> The nexus of postwar terrorist operations, then, can be traced to Skorzeny's branch of the Nazi International, and its deep connections both to radical Islamicism, and to the elite of the American corporate world and rogue groups within American intelligence.[38]

Farrell goes on to say that Skorzeny and his branch got into the international drug trade as well, possibly using their flying saucers to move heroin and cocaine around the world. He points out that Skorzeny and his Nazi International had no government oversight or scrutiny. They could basically operate as they saw fit and go anywhere they wanted to. Says Farrell, referencing the book *The Great Heroin Coup: Drugs, Intelligence and International Fascism*,[43] for his information:

> This story begins in the 1970s, when a major effort was mounted by rogue elements within American intelligence, and by the postwar Nazi International, to restructure the entire global heroin trade, closing off the old "French Connection" through Marseilles, and re-orienting it through the Southeast Asian, Mexican, and Latin American drug cartels. Placing the lucrative Latin American and Mexican cartels' trade in league with the postwar Latin American Nazis and Fascists gave this component of the breakaway civilization a virtually inexhaustible source of money to fund its activities, from terrorism to research.
>
> This group had pulled off a number of spectacular

Otto Skorzeny on left with Juan Perón, center, circa 1950.

215

robberies in Europe, including the theft of gold bullion from the Société Générale de Nice, had been implicated in various assassination attempts on French President Charles DeGaulle, and had connections to terrorist and Fascist groups in Italy, Lebanon, and Britain. So pervasive and deep were these connections that "a number of newspapers, even including the *New York Times*, mentioned speculations" that it might be the activity of a "Fascist International."[38]

Farrell says that the Nazi component of the breakaway civilization moved quickly to solidify its ties and influence over the international drug trade. As this drug trade was being restructured in the early 1970s, Miami became a haven for the activity of European Fascists, he says and then gives us this informative quote from Henrik Kruger's *The Great Heroin Coup: Drugs, Intelligence and International Fascism*:

> What are European Fascists doing in Miami before and after major operations? Why are Miami-based Cuban exiles executing contracts on young Spaniards? Why was the main station of the CIA-supported Fascist front World Service in Miami? Why did bank robber Spaggiari contact the CIA in the United States?
>
> Miami is the center of a huge conspiratorial milieu whose personnel wind through the Bay of Pigs, attempts on Castro's life, the JFK murder and the great heroin coup, and which is now reaching out with a vengeance to Latin America and Europe.
>
> To trace the roots of this milieu we must refer to the immediate aftermath of World War II, when the CIA began its close cooperation with Adolf Hitler's espionage chief, Reinhard Gehlen, and the Soviet general, Andrei Vlassov, of Russia's secret anti-Communist spy network. Vlassov's organization was absorbed into Gehlen's, which evolved into a European subsidiary of the CIA. U.S. and German agents mingled in Berlin and West Germany, paving the way for inroads into U.S. intelligence by former Nazis, SS agents, and Russian czarists.

216

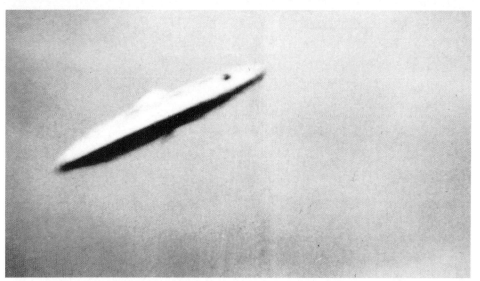

This photograph was taken by Rainer Weiss in St. Georgen, Germany in November, 1978.

Left: A Vril saucer photographed over Obihiro-Hokkaido, Japan on November 12, 1974.

This series of four photos was taken in March 1967 in Yungay, Peru. In the first two photos, two disks approach the shoulder of a mountain and then split up. One circles around the valley below.

These final two photos taken in March 1967 in Yungay, Peru show one of the disks turning toward the photographer and meeting with the second disk; the two disks then fly away into the clouds.

Left: A Vril saucer photographed in flight, date unknown.

The Revell model of a Haunebu II.

An SS Death Head (Totenkopf) ring and special ring box with the SS runes on it. As the sign of a secret society, it was to be worn on the third finger of the left hand.

Top left to bottom left: In March of 1971 the US Navy submarine *Trepang* took these photos of a cigar-shaped craft coming up out of the ocean near Iceland. The displaced water turned to steam as the craft leveled and flew off. A second smaller object then came out of the water.

Left and below: Rare color photos of the Andromeda craft in flight. These photos were part of the 1989 document dump to Ralf Ettl. These craft were submarines and aircraft. How many were built after the war we simply don't know.

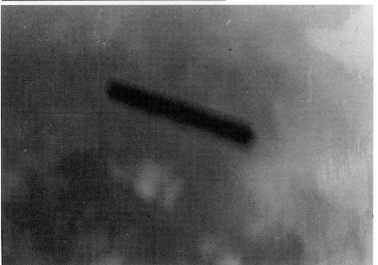

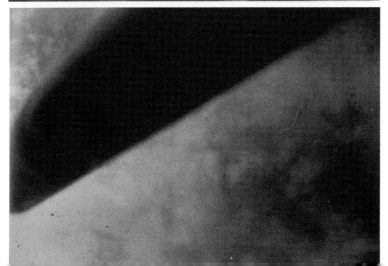

A photo of a saucer over a mining area in northern Chile taken on April 30, 2013.

The photo of a saucer craft coming suddenly out of the water taken by seal hunters along the coast of the Peninsula de Valdez in Chubut Province, Patagonia, Argentina Dec. 2, 1971.

Above: A screen shot of a Haunebu at Cerro Uritorco from a television newscast. *Right*: The 1976 Argentine book *Historia de los Platos Voladores en la Argentina* (*The History of Flying Saucers in Argentina*). This book has never been published in English.

A photo of a flying saucer over Cerro Uritorco in northern Argentina.

Headquarters of the CIA/Gehlen/Vlassov combine, staffed in the mid-fifties by 4000 full-time agents, were in Pullach, near Munich. There Gehlen sang to the tune of more than one piper, having remained in touch with the old Nazi hierarchy relocated in Latin America, whose coordinator, Otto Skorzeny, was in Spain. Skorzeny had infiltrated the Spanish intelligence agency DGS, and effectively controlled it single-handedly.

With the onset of the Cold War, Gehlen's agents were recruited by the CIA for assignments in the United States, Latin America and Africa. One agent, reportedly, was Frank Bender, allegedly alias Frank Swend, a key figure in the Bay of Pigs invasion.[43]

So we have a worldwide network of fascists with offices in Germany, Spain, the US, Argentina and all over Latin America. A network of banks, money, factories, intelligence agents, and let us not forget, Haunebu craft. After all, someone has to be flying around in these disks and cigar-shaped craft. What were these secret missions being made by the craft? Were some of them carrying drugs to Europe and the US? The secret cities of the SS, the factories that had been making the craft, including the flying

A Vril craft over Trowbridge, Wiltshire, UK taken by John Wicks, Aug. 1974.

submarines, must have been closed at some point in the late 1950s or 1960s. Or, were they now being jointly run with a wayward faction of the CIA? Both theories have been offered by German writers who claim to have knowledge of these bases. Still, the factories and bases in Argentina would continue to function, and even secret cities of the SS in Chile and Paraguay. As I have said earlier, it seems that the Germans—controlled by the SS—were doing a wide variety of major projects, including underground tunneling, in Peru, Venezuela, Bolivia and elsewhere during and after the war. They were apparently funded by Bormann and directed by Skorzeny and Hans-Ulrich Udel. Whole factories in Cordoba, Argentina were part of the complex web of money, technology and industry. The Haunebu factory north of Cordoba appears to function even to this day.

Chapter 6

Set the Controls for the Heart of the Sun

Over the mountain watching the watcher
Breaking the darkness waking the grapevine
Knowledge of love is knowledge of shadow
Love is the shadow that ripens the vine
Set the controls for the heart of the sun
—*Set the Controls for the Heart of the Sun*, Pink Floyd

Knowing that the Haunebu and Vril craft were flying around the world for decades, one would think that there would be evidence of such activity such as photos, documents and testimonials. Indeed, there is. Much comes from South America which has always been a hotbed of UFO activity and still is.

UFO Sightings in Brazil
The first known UFO sighting in Brazil was on July 23, 1947, when a topographer named José Higgins was working with many laborers near Pitanga, a suburb of Bauru, in São Paulo state. Suddenly, the group heard an extremely sharp sound. Some moments later, they saw a lens-shaped object landing near them. According to Wikipedia, the workmen ran away in fright, leaving Higgins alone. Higgins reported that three humanoid figures, he thought about seven feet tall, dressed in inflated, transparent suits, emerged and spoke to him in an unknown language.

They tried to get Higgins to enter their craft, but the witness refused, and they allowed him to go. The beings remained in the area for a half hour, observed by the witness from some distance

219

away, then re-entered the object, which took off. The case was reported by the Lorenzens in the APRO Bulletin of May of 1961.

Here is a curious case that may involve a Haunebu or Vril craft. The Lorenzens and other UFO investigators may find this to be evidence of extraterrestrials visiting southern Brazil, in the vicinity of one of Brazil's largest cities, looking for earthmen to take a ride with them in their flying saucer. One wonders what would have happened to Higgins if he had entered the craft. These friendly UFOnauts might have taken him to a base in Antarctica. And what language were they speaking? Why would extraterrestrials speak to someone in their alien language and expect someone from Earth to understand them? Were they actually speaking some Earthly language such as Volga German or the Uzbeck language? Maybe even Tibetan? Romanian and Ukrainian are two other possible languages that might be spoken by former SS operatives immediately after the war. We are not told what languages Higgins spoke and understood, but he may only have spoken Portuguese and possibly some English.

Five years later on May 5, 1952, the journalist Joao Martins and the photographer Eduardo Keffel claimed to have seen a flying disk in the vicinity of Barra da Tijuca. Keffel took at least five photographs of a flying disk, which were published by the magazine *O Cruzeiro*. Barra da Tijuca is a neighborhood or bairro in Rio de Janeiro, located in the southwestern portion of the city on the Atlantic Ocean along Barra Beach. Barra is well known for its many lakes and rivers, and its beach and party lifestyle.

They say that a picture is worth a thousand words, and it is the photos that Keffel took which are quite startling. The photos, one shown here, are taken in the daylight and the flying saucer, with a dome on top, is very clearly seen in flight. This is not your "light in the sky" UFO photo but one that is very clear and unambiguous. It appears to be a modified Haunebu. One has to wonder whether this photo session was planned with the journalist and photographer, as the object seems to want to be photographed. This is a curious aspect of many of the early UFO cases in South America in the later 1940s, 1950s and 1960s, the UFOs seem to be making an effort to be seen by large numbers of people, such as the UFO sightings over Buenos Aires or that photographed by

A photo taken at the Rio de Janeiro beach of Barra de Tijuca on
May 5, 1952.

Keffel and published in *O Cruzeiro*. Yes, these flying saucers want publicity and to be photographed. It's a bit like waving one's arms and shouting, "Hey, look over here!"

Keffel's photos are either fakes, or they show a genuine flying disk in flight. Was this a Haunebu built in Germany or had it been built at the so-called Haunebu factory at Beaver Dam in Greenland?

In 1951 Coral Lorenzen moved with her Air Force husband to Burbank, California where friends of hers told her that she should form an organization devoted to studying UFO phenomena. She and her husband moved to Green Bay, Wisconsin and it was here that she started APRO in January 1952 and published the first mimeographed newsletter in July of that year. Coral also began working as a journalist for the *Green Bay Press-Gazette*.

Coral says that she and her husband made a trip to Los Angeles in 1956 to visit some friends and were told of a former Navy pilot who had encountered two flying disks over Korea in September of 1950. They visited the pilot, whom they call "Mr. Douglas" and he told them the story of how his plane and two other fighter-bombers took off from the deck of an aircraft carrier off the Korean coast and were flying north up a valley looking for a North Korean convoy of trucks that they were to strafe and bomb. It was just before sunrise and as the sun rose he and the other planes scoured the mountains and the valley for the convoy that they were to target. Says Lorenzen in her book, quoting "Mr. Douglas":

> I was watching the ground below for the convoy and I was startled to see two circular shadows coming along the ground from the northwest at a high rate of speed. We were flying north above a valley which was surrounded on the east and west by mountains, with a pass directly ahead of us to the north. When I saw the shadows I looked up and saw the objects that were creating them. They were huge. I knew that as soon as I looked at my radar screen. They were also going at a good clip—about 1000 or 1200 miles per hour. My radar display indicated one and a half miles between the objects and our planes when the objects suddenly seemed to halt, back up and begin a 'jittering' or

'fibrillating motion.' My first reaction, of course, was to shoot. I readied my guns, which automatically readied the gun cameras. When I readied the guns, however, the radar went haywire. The screen 'bloomed' and became very bright. I tried to reduce the brightness by turning down the sensitivity, but this had no effect. I realized my radar had been jammed and was useless. I then called the carrier, using the code name. I said the code name twice, and my receiver was out—blocked by a strange buzzing noise. I tried two other frequencies, but could not get through. Each time I switched frequencies the band was clear for moment, then the buzzing began.

While this was going on the objects were still jittering out there ahead of us, maintaining our speed. About the time I gave up trying to radio the carrier the things began maneuvering around our planes, circling above and below. I got a good look at them. I had never seen anything like them before, and I learned after we reached the carrier that the other men that flight were of the same opinion. They were huge—I said that before. Before my radar set was put out of commission, I used the indicated range plus points of reference on the canopy to determine their size. They were at least 600 or possibly 700 feet in diameter.

The objects had a 'silvered mirror' appearance, with a reddish glow surrounding them. They were shaped somewhat like a coolie's hat, with oblong ports from which emanated a copper-green colored light which gradually shifted to pale pastel-colored lights and back to the copper-green again. Above the ports was a shimmering red ring which encircled the top portion.

When the things maneuvered above us, we saw the bottoms of them. In the middle of the underside was a circular area, coal black and nonreflective. It was simply inky black and it is important to note that although the whole object 'jittered' while maneuvering, the black circular portion on the bottom was steady and showed no indication of movement.[17]

Coral and her husband peppered the pilot with questions and he said that the two disks suddenly took off in the direction for which they had come at a high rate of speed, and the three Navy jets returned to the aircraft carrier. They were debriefed individually and they and their jets were tested for radiation. They learned that some of their instrument dials had become luminous and that all of the gun camera film had been fogged or exposed.

"Mr. Douglas" then said to Coral: "Mrs. Lorenzen, those things were real—intelligently controlled machines."

Coral Lorenzen then asked the pilot: "Where do you think they could have come from?"

The pilot answered her, "I would be presuming a lot if I were to try to definitively give their origin, but I do know this—those things were the products of intelligences far beyond ours—a science far in advance of ours. They must be from outer space. There's no other answer." And so the second chapter of her book *Flying Saucers*[17] ends, and in fact the title of this chapter is "No Other Answer."

As one reads Lorenzen's book, one with many excellent and curious UFO encounters, one quickly realizes that she is completely convinced that flying saucers and cigar-shaped craft are real and that they must be from another planet. She quickly theorizes that the craft are actually from Mars and are piloted by diminutive Martians. The range of foo fighter activity and German secret disk programs was completely unknown to Lorenzen at the time she wrote her book (the 1950s), as were the secret German bases in Antarctica, Argentina and elsewhere. It wasn't really until the 1970s and 80s that knowledge of the German saucer programs was leaked to the public, and eventually CIA documents obtained through the Freedom of Information Act showed that the CIA, along with the FBI and the military, were aware of this activity and that it was continuing in South America and elsewhere.

Interestingly, Lorenzen's book contains mainly UFO reports from South America. This is because she had a number of devoted APRO members in several South American countries who sent her material from this mysterious continent, a continent with a great deal of UFO activity.

Lorenzen gives us a small collection of her first APRO

catalogued incidents which are from 1952. She discusses a couple of other curious incidents over Japan and Korea in March of 1952 and then some sightings in the Baltic States like Latvia, Finland and Sweden, plus the "French flap" of 1952 and finally the famous Washington D.C. influx of formations of flying saucers over the Capitol spanning several days and nights in from July 12 to July 29, 1952.

Wikipedia tells us about this famous encounter:

At 11:40 p.m. on Saturday, July 19, 1952, Edward Nugent, an air traffic controller at Washington National Airport (today Ronald Reagan Washington National Airport), spotted seven objects on his radar. The objects were located 15 miles (24 km) south-southwest of the city; no known aircraft were in the area and the objects were not following any established flight paths. Nugent's superior, Harry Barnes, a senior air-traffic controller at the airport, watched the objects on Nugent's radarscope. He later wrote:

We knew immediately that a very strange situation existed... their movements were completely radical compared to those of ordinary aircraft.

Barnes had two controllers check Nugent's radar; they found that it was working normally. Barnes then called National Airport's radar-equipped control tower; the controllers there, Howard Cocklin and Joe Zacko, said that they also had unidentified blips on their radar screen, and that they had seen "a bright light hovering in the sky... [it] took off, zooming away at incredible speed." Cocklin asked Zacko, "Did you see that? What the hell was that?"

At this point, other objects appeared in all sectors of the radarscope; when they moved over the White House and the United States Capitol, Barnes called Andrews Air Force Base, located 10 miles from National Airport. Although Andrews reported that they had no unusual objects on their radar, an airman soon called the base's control tower

225

A photo of some of the UFO activity around Washington D.C. in 1952.

to report the sighting of a strange object. Airman William Brady, who was in the tower, then saw an "object which appeared to be like an orange ball of fire, trailing a tail... [it was] unlike anything I had ever seen before." As Brady tried to alert the other personnel in the tower, the strange object "took off at an unbelievable speed."

On one of National Airport's runways, S.C. Pierman, a Capital Airlines pilot, was waiting in the cockpit of his DC-4 for permission to take off. After spotting what he believed to be a meteor, he was told that the control tower's radar had detected unknown objects closing in on his position. Pierman observed six objects—"white, tailless, fast-moving lights"—over a 14-minute period. Pierman was in radio contact with Barnes during his sighting, and Barnes later related that "each sighting coincided with a pip we could see near his plane. When he reported that the light streaked off at a high speed, it disappeared on our scope."

Meanwhile, at Andrews Air Force Base, the control tower personnel were tracking on radar what some thought to be unknown objects, but others suspected, and in one instance were able to prove, were simply stars and meteors. However, Staff Sgt. Charles Davenport observed an orange-red light to the south; the light "would appear to stand still, then make an abrupt change in direction and altitude… this happened several times." At one point both radar centers at National Airport and the radar at Andrews Air Force Base were tracking an object hovering over a radio beacon. The object vanished in all three radar centers at the same time.

At 3 a.m., shortly before two United States Air Force F-94 Starfire jet fighters from New Castle Air Force Base in Delaware arrived over Washington, all of the objects vanished from the radar at National Airport. However, when the jets ran low on fuel and left, the objects returned, which convinced Barnes that "the UFOs were monitoring radio traffic and behaving accordingly." The objects were last detected by radar at 5:30 a.m.

The sightings of July 19–20, 1952, made front-page headlines in newspapers around the nation. A typical example was the headline from the *Cedar Rapids Gazette* in Iowa. It read "SAUCERS SWARM OVER CAPITAL"

in large black type. By coincidence, USAF Captain Edward J. Ruppelt, the supervisor of the Air Force's Project Blue Book investigation into UFO sightings, was in Washington at the time. However, he did not learn about the sightings until Monday, July 21, when he read the headlines in a Washington-area newspaper. After talking with intelligence officers at the Pentagon about the sightings, Ruppelt spent several hours trying to obtain a staff car so he could travel around Washington to investigate the sightings, but was refused as only generals and senior colonels could use staff cars. He was told that he could rent a taxicab with his own money; by this point Ruppelt was so frustrated that he left Washington and flew back to Blue Book's headquarters at Wright-Patterson AFB in Dayton, Ohio. Upon returning to Dayton, Ruppelt spoke with an Air Force radar specialist, Captain Roy James, who felt that unusual weather conditions could have caused the unknown radar targets.

At 8:15 p.m. on Saturday, July 26, 1952, a pilot and stewardess on a National Airlines flight into Washington observed some lights above their plane. Within minutes, both radar centers at National Airport, and the radar at Andrews AFB, were tracking more unknown objects. USAF master sergeant Charles E. Cummings visually observed the objects at Andrews; he later said that "these lights did not have the characteristics of shooting stars. There was [sic] no trails... they traveled faster than any shooting star I have ever seen." Meanwhile, Albert M. Chop, the press spokesman for Project Blue Book, arrived at National Airport and, due to security concerns, denied several reporters' requests to photograph the radar screens. He then joined the radar center personnel. By this time (9:30 p.m.) the radar center was detecting unknown objects in every sector. At times the objects traveled slowly; at other times they reversed direction and moved across the radarscope at speeds calculated at up to 7,000 mph (11,250 km/h). At 11:30 p.m., two U.S. Air Force F-94 Starfire jet fighters from New Castle Air Force Base in Delaware arrived over Washington. Captain John

McHugo, the flight leader, was vectored towards the radar blips but saw nothing, despite repeated attempts. However, his wingman, Lieutenant William Patterson, did see four white "glows" and chased them. He later said that, "I tried to make contact with the bogies below 1,000 feet. I was at my maximum speed... I ceased chasing them because I saw no chance of overtaking them." According to Albert Chop, when ground control asked Patterson "if he saw anything," Patterson replied "'I see them now and they're all around me. What should I do?' ...And nobody answered, because we didn't know what to tell him."

The Robertson Panel and UFO Hysteria

There were good photographs taken of some of the incidents, and one of the photos, seen here, clearly shows nine glowing circular objects around the Capitol building. This seems to have been a Haunebu or Vril flyover, a show of force in a sense, to folks in Washington D.C. The Air Force decided to explain the incidents away as temperature inversions and problems with early radar that could also pick up flocks of birds. However, this incident led to the little-known Robertson Panel which was to recommend that the Air Force not spend any time discussing UFO incidents that they could not easily explain as there was a certain "mass hysteria" that was taking place concerning the public and flying saucers. Says Wikipedia of the Robertson Panel:

> The extremely high numbers of UFO reports in 1952 disturbed both the Air Force and the Central Intelligence Agency (CIA). Both groups felt that an enemy nation could deliberately flood the U.S. with false UFO reports, causing mass panic and allowing them to launch a sneak attack. On September 24, 1952, the CIA's Office of Scientific Intelligence (OSI) sent a memorandum to Walter B. Smith, the CIA's Director. The memo stated that "the flying saucer situation... [has] national security implications... [in] the public concern with the phenomena... lies the potential for the touching-off of mass hysteria and panic." The result of this memorandum was the creation in January 1953 of

the Robertson Panel. Dr. Howard P. Robertson, a physicist, chaired the panel, which consisted of prominent scientists and which spent four days examining the "best" UFO cases collected by Project Blue Book. The panel dismissed nearly all of the UFO cases it examined as not representing anything unusual or threatening to national security. In the panel's controversial estimate, the Air Force and Project Blue Book needed to spend less time analyzing and studying UFO reports and more time publicly debunking them. The panel recommended that the Air Force and Project Blue Book should take steps to "strip the Unidentified Flying Objects of the special status they have been given and the aura of mystery they have unfortunately acquired." Following the panel's recommendation, Project Blue Book would rarely publicize any UFO case that it had not labeled as "solved"; unsolved cases were rarely mentioned by the Air Force.

Coral Lorenzen and her APRO group were well aware of the July Washington D.C. UFO incidents and covered them extensively in her newsletter. However, she knew nothing of the Robertson Panel, and such government panels and documents concerning UFOs would not be declassified until many decades later.

What seems to have happened with the July 1952 swarm over Washington D.C. was exactly the kind of "UFO psyops" from a hostile power that the Robertson Panel was warning about: an overt program over several days to highlight UFOs in the national press of the USA and around the world. This kind of activity is "hey, look at me" sort of behavior. They want to be seen, they want to cause a ruckus, they want headlines in the newspapers. They got it. Aliens coming to planet Earth to steal our DNA or planet do not need to fly in swarms over Washington D.C. Supposedly they do this activity in secret. So why the "look at me" activity?

This "fleet" of flying saucers, coming in from the Atlantic Ocean, was apparently a group of Haunebu and Vril saucers that were making very calculated maneuvers over Washington D.C. in a highly coordinated manner. If this activity originated from the secret SS bases in South America, Antarctica, Spanish Sahara

230

CENTRAL INTELLIGENCE AGENCY
WASHINGTON 25, D. C.

OFFICE OF THE DIRECTOR

1952

MEMORANDUM TO: Director, Psychological Strategy Board

SUBJECT: Flying Saucers

1. I am today transmitting to the National Security Council a proposal (TAB A) in which it is concluded that the problems connected with unidentified flying objects appear to have implications for psychological warfare as well as for intelligence and operations.

2. The background for this view is presented in some detail in TAB B.

3. I suggest that we discuss at an early board meeting the possible offensive or defensive utilization of these phenomena for psychological warfare purposes.

Enclosure

Walter B. Smith
Director

A 1952 CIA memo discussing flying saucers as psychological warfare.

and Europe, perhaps this Nazi remnant did it as a show of power to the US military. At this time, Reinhard Gehlen, the former SS chief, was working with the Americans and Operation Paperclip had begun over five years before.

The range of a Haunebu or Vril craft was "pole to pole," and they could have come from a base in Antarctica. As I have discussed earlier and in other books, they may have flown to Washington D.C. from Greenland, Norway, the Spanish Sahara or just as easily from anywhere in South America. The speed and range of these craft made nearly any mission possible.

Back in Brazil, a very curious UFO incident happened on the evening of November 4, 1957 at a Brazilian military fort. At the Itaipu Fort in Praia Grande, São Paulo state, two sentinels suffered moderate burns after being hit by a heat wave from a flying saucer that descended on the fort from the sky. During the incident the entire electrical system of the fort, including the emergency circuits, went down and the fort was blacked out

231

In the wake of this seeming attack on a Brazilian military base by a flying saucer, Brazilian Army, along with investigators of the Brazilian Air Force, flew to the fort to interview the soldiers. Also with them were United States Air Force (USAF) personnel who were sent from the US to investigate the incident. No explanation was ever given.

Author, investigator and US Air Force officer Donald Keyhoe expressed his opinions on this bizarre case of an extraterrestrial flying saucer attacking a Brazilian fort:

> Such civilization could see that in Earth we now have atomic bombs and that we are quickly improving our rockets. Given the past history of mankind—frequent wars showing a belligerent human race—they must have become alarmed. We should, therefore, expect, especially at these times, to receive such visits. According to this, the main objective of the aliens would be to watch our space improvements, fearing we could become a threat to other planets. If this hypothesis is exact, it could be expanded to link the launching of the Sputniks with the attack to the Fort Itaipu. However, this sounded absurd for all investigators. It would mean that the aliens would be worried about our firsts steps in space, and by small space ships so primitive that would look like a canoe if compared to a transatlantic liner. It would also mean that those burnings had the purpose of demonstrating the superior weapons they could use against the aggressive explorers coming from the Earth. However, we were still far from piloted space flight, even to the Moon. According to human logic, we would not be able to threaten a superior space ship—not now nor later.

So for Keyhoe, the idea was that extraterrestrials with their advanced technology would not be afraid of us or our technology, and would therefore not have attacked the Itaipu Fort. Keyhoe stops his musings there, but if he were to continue he would wonder what earthly power would have attacked the Brazilian fort? Depending on what he may have known at the time he might have been aware of the German flying saucers and the Nazi base in Antarctica. For here is an earthly power that would have reasons to attack a Brazilian fort.

The reasons could be numerous. Brazil had declared war against Germany in the last years of the war and maintained a joint submarine base with the Americans on Trindade Island, 600 miles out into the Atlantic off the coast of Brazil. Most of the countries in Latin America were neutral during WWII, but Brazil was persuaded by the Americans to join with the Allies and conduct joint naval exercises. This island would have its own flying saucer encounter only two months later, to be discussed shortly. The attack on the fort, taking the power supply out and zapping a few sentinels with a heat wave, may have just been a warning attack on a Brazilian facility to let them know that they were dealing with an independent power—either extraterrestrials or the Nazi International and their Haunebu and Vril craft. It seems likely that the attack on Itaipu Fort was done with a Vril Jager 7, but no one can know for sure.

The 1958 Incidents at Trindade Island

During this same time were the famous daylight sightings of a large flying saucer seen by dozens at Trindade Island in the mid-Atlantic starting in November of 1957 and culminating on January 16, 1958.

Trindade is a small rocky island in the middle of the South Atlantic Ocean between the Brazilian coast and the African continent, more than six hundred miles off the coast of Bahia and Espírito Santo. It was used as an American and Brazilian anti-submarine base during World War II. Abandoned after the war, it was reoccupied in October of 1957 (just before the attack on Itaipu Fort), when a Brazilian Navy task force built an oceanographic post and meteorological station there.

This was part of the International Geophysical Year research program, and during November of 1957 instrument-carrying meteorological balloons were released daily to study high-atmosphere conditions. During one flight in late November a weather balloon was launched and a shiny disk-shaped object was seen in the sky near it. The balloon burst at the proper time and the unconventional object stayed in sight for three hours, eventually disappearing upward and out of sight.

There was a full moon at the time and a second sighting took place on December 25, 1957, when a laborer on the island described an object he had seen as being silvery in color, round, with an angular diameter comparable to that of the full moon and flying silently at great height. A third sighting came on December 31 when a similar object passed over the island again at 7:50 AM. The silvery, circular object, with the apparent size of the full moon, crossed the sky silently at about 6,000 feet.

On the following day, January 1, 1958, at 7:50 AM a bright light flashed over the sea at high speed. The flying saucer suddenly turned at a 90-degree angle and then vanished into the horizon. The object was viewed by the whole garrison, and workers and sailors reported that it was the same object they had sighted on previous occasions.

The next day (January 2) saw a fifth sighting where the crew of a ship witnessed a discoid craft with a weird orange glow maneuver at high speed and made sudden changes of course with right-angle turns. At times it briefly hovered motionless, sometimes close to the ship. The flying saucer seemed to be doing maneuvers to impress the crew and show them what the craft was capable of.

Then on January 6 a released weather balloon was being tracked from the ground. The sky was blue and clear with no haze, and there was only a solitary cumulus cloud almost overhead. The balloon's transmitter became silent. Commander Bacellar went outside to investigate and things appeared normal—the balloon was high in the sky and still climbing, slowly approaching the solitary cumulus cloud overhead at fourteen thousand feet, the height at which the balloon's instruments were to be jettisoned.

Suddenly, the balloon appeared to be sucked upward into the cloud and was lost to the viewer's sight. The balloon reappeared ten minutes later and resumed its ascent, more rapidly now, but it no longer had its instruments. The balloon had gone into the cloud with its full instrument load and had reappeared without it. The instruments were never found and observers did not see them come down.

Then, shortly after the balloon reappeared another object left the cloud. A silvery disk the color of polished aluminum came slowly from behind the cloud, moving from a southwest direction to the southeast.

A seventh sighting reportedly took place around January 10. The flying disk came in very low over the island, flashed toward the meteorological post at high speed, slowed abruptly and hovered for a brief moment above Desejado Peak, then moved again in a zigzag course and was gone into the horizon at tremendous speed. The object was described as resembling a highly polished flattened spheroid with a large ring circling its equator. The ring appeared to be rotating at high speed, and the objet made no sound as it flew. It was surrounded by an eerie greenish glow that almost disappeared while the object was hovering and became brighter when the object moved.

Witnesses described the object as two or three times the size of a DC-3, and they said it appeared to be intelligently controlled.

Then at 12:00 PM on January 16, 1958, the Brazilian ship *Almirante Saldanha*, one of many ships in the task force, was preparing to sail away from Ilha de Trindade. Captain Viegas was on the deck with several scientists and members of the crew when he suddenly noticed a flying object, which had a "ring" around it, just like Saturn.

Everyone reportedly saw the UFO at the same time. It came to the island from the east, flew towards the Pico Desejado (Wished Peak), made a turn and went away very quickly to the northwest. As soon as the object was noticed Almiro Barauna was requested to get a camera and photograph the flying object. After getting the camera and going up on deck, he stated that he managed to take a number of photos of the object.

Almiro Barauna was a member of the Icarai Club for Submarine Hunting and had been invited aboard the *Almirante Saldanha* by the Navy to take underwater photographs. At the time the object was sighted, Barauna was able to get four good exposures of the object. He later made the following statement to the Rio de Janeiro newspaper *Jornal Do Brasil*:

Suddenly Mr. Amilar Vieira and Captain Viegas called me, pointing to a certain spot in the sky and yelling about a bright object which was approaching the island. At this same moment, when I was still trying to see what it was, Lieutenant Homero, the ship's dentist came from the bow toward us, running, pointing to the sky and yelling about the object he was watching. Then I was finally able to locate the object by the flash it emitted. It was already close to the island. It glittered at times, perhaps reflecting sunlight, perhaps changing its own light—I don't know. It was coming over the sea, moving toward the point called the "Galo Crest." I had lost about thirty seconds looking for the object, but the camera was already in my hands, ready, when I sighted it clearly silhouetted against the clouds. I shot two photos before it disappeared behind Desejado peak. My camera was set at a speed of .125 with an f/8 aperture, and this was the cause of the overexposure error, as I discovered later. The object remained out of sight for a few seconds, behind the peak, reappearing larger and closer than before and moving at a higher speed. I shot the third photo. The fourth and fifth shots were lost, partly because of the speed at which the object was moving, and partly because I was being pushed and pulled about in the excitement. It was moving in the direction from which it had come, and it appeared to stop in mid-air for a brief time. At that moment I shot my last photo, the last on the film. After about ten seconds the object continued to increase its distance from the ship, gradually diminishing in size and finally disappearing into the horizon.

Barauna told the newspaper that the object made no discernible sound; that it was dark gray in color, appearing to be surrounded, mostly in the area ahead of it, by a kind of condensation of a greenish, phosphorescent vapor or mist; it definitely appeared to be a solid object; and its flight was an undulatory movement across the sky—"like the flight of a bat." In the minutes after the sighting, Barauna said, "The ship's commander and several officers from the garrison wanted to see what I had got in the photos. As I was curious, too, I decided to develop the exposed film at once, aboard the ship. The processing was done under the supervision of several officers, including Commander Carlos A. Bacellar. There was no photographic paper on the ship, so only the negatives were seen while still aboard. They were seen and examined by the whole crew, however."

Two days after the return of the ship to Rio de Janeiro, Commander Bacellar came to Barauna's residence and asked if he could see the enlargements made from the negatives. He took them to the Navy authorities and returned them to the photographer two days later, and congratulated Barauna. Then Barauna was asked to appear at the Navy Ministry, where he was questioned by high-ranking staff officers. At the meeting the officers asked for the negatives. They were sent to the Cruzeiro do Sul Aerophotogrammetric Service where they were analyzed, and Barauna was told the negatives were found to be genuine, definitely excluding the possibility of a trick or falsification.

On a second visit to the Navy, tests were performed and Barauna worked with his Rolleiflex, taking shots at the same time intervals he had used to photograph the object, while three officers with stopwatches recorded the time intervals. Based on charts of the island they concluded that the object was flying between 900 and 1000 kilometers per hour (about 600-700 miles per hour), and its size was determined to be about 120 feet in diameter and about 24 feet high.

Barauna later said that he had seen a thick dossier that the officers consulted constantly during their interrogation of him. He was informed that his pictures, mixed with others, had been shown to witnesses for identification. His photos were selected as those which caught the object seen at Trindade.

Barauna was quoted in *O Globo*: "At the end of the meeting, the chief intelligence officer said he was convinced my photos were authentic. He showed me another photo that had been taken by a Navy telegrapher sergeant, also at Trindade. A box camera had been used. The photo showed the same object seen in my pictures. They told me it had been taken before my arrival at Trindade Island."

The Rio de Janeiro newspaper *Ultima Hora* roported on February 21, 1958 that at least a hundred individuals had witnessed the sighting of the object and that the four Barauna exposures were

The photo sold to the media that was allegedly taken at Ilha de Trindade, 1958.

obtained within fourteen seconds. Only one photograph was ever released to the press by Barauna, and this photo was deemed a hoax by some researchers. Then—get this— Barauna apparently admitted in 2010 that he had created the photo that was sold to the newspapers, seen here.

Since this case is one of the most famous UFO incidents in Brazil, when a female friend of Barauna named Emilia Bittencourt spoke on a television shown in 2010 and claimed that Barauna had faked the famous photo published in the newspapers, Brazilians were stunned. Bittencourt claimed that Barauna had created a montage:

> He got two spoons, joined them and improvised a spaceship, using as a background his home fridge. He photographed on the fridge door an object with perfect lighting, because he calculated everything, he wasn't dumb. He laughed a lot.

So, as a result of this confession it is now claimed that the whole episode was a hoax. What? Hundreds of people were said to have seen the flying saucer the day it was photographed and the UFO activity had been going on there since November. Barauna had supposedly taken at least four good photos and claimed that the Brazilian Navy officials had a thick dossier about the incident and other photos. It is impossible that Barauna faked this entire incident. Clearly something else is going on here.

The photo that Almiro Barauna sold to the newspapers in Rio de Janeiro, Sao Paulo, and other cities, was quite possibly a fake. He may well have faked it just as he is described doing, using his fridge door as a background. The reason he would have done this is that the Brazilian Navy apparently took his photos and negatives and kept them for themselves. Left without any photos to sell to the newspapers he decided to fake a photo, knowing what the originals looked like. He then sold this photo and his story to all the newspapers he could, all in a photographer's quest for money.

In this case it would seem that what unfolded at this remote ex-submarine base over two months is pretty remarkable:

1) A former war base is reactivated. As soon as men and equipment are moved in, unconventional aerial objects are seen in the area.
2) There is an indication that a flying saucer interfered with meteorological equipment and possibly absconded with one of the meteorological devices.
3) The Brazilian Navy attempted to suppress information on the many sightings.
4) The press helped pressure the authorities to release more information.
5) While photos were taken, they are apparently missing and the one photo available is supposed to be a fake.

What seems to be going on here is that Haunebu or Vril craft are flying around this remote Atlantic island and are deliberately showing themselves to the many crew and workers now suddenly at the island. They deliberately seize one of the meteorological devices, either to look at the technology, or just to do something that an entire ship's crew might see—in broad daylight no less. As in many cases in South America, the craft make little effort to disguise themselves and instead perform aerial acrobatics for everyone's amusement. Basically, they are trying to show as many people as they can, including Brazilian Naval personnel, that they are here and what they can do, which is pretty fantastic from a flight standpoint.

One also has to wonder if this former submarine base was being used by someone else in the years after the war. This event took place some thirteen years after the war officially ended and it seems that the base was probably closed shortly after the war. Did the SS then use the submarine base at Trindade Island for its own purposes? This island is remote enough from Brazil that all sorts of activity could have gone on there starting in 1947. U-boats still active from the bases in Antarctica, the Canary Islands and Greenland could have also used the decommissioned submarine base at Trindade.

Was this the reason that Itaipu Fort in Praia Grande, São Paulo state was attacked by a flying saucer? Was it a warning to the Brazilian Navy not to interfere at the ex-submarine base at Trindade? The curious attack on this military fort is one of the few incidents of its type ever recorded and doesn't make a lot of sense, except as an attack by an SS crew in a Vril or Haunebu on a Brazilian military fort. No attacks like this have ever been recorded in Argentina or other South American countries with a large ex-Nazi presence.

Once more, the Third Power is able to flex its muscle and show just how far the tentacles of this octopus can reach. One has to wonder if there was a Nazi U-boat in the vicinity of Trindade during the sightings of 1957-1958? Also, where was the Haunebu coming from—Antarctica? It may have been coming from a Haunebu base in Patagonia as well. This case is still famous in Brazil as one of the earliest mass sightings.

A Haunebu Lands in Melbourne, Australia in 1966

Another famous incident of a flying saucer getting lots of attention is the 1966 Westall High School sighting in Melbourne, Australia. This UFO incident happened in broad daylight in a suburb of Melbourne, Australia. On April 6 of that year, around 11:00 AM, for about 20 minutes, more than 200 students and teachers at two Victoria state schools allegedly witnessed an unexplained flying object—looking like a flying saucer—which

An alleged photo of the Westall Haunebu encounter in Melbourne, Australia.

descended into a nearby open wild grass field. The grass field was adjacent to a grove of pine trees in an area known as The Grange (now a nature reserve). According to reports, the object then ascended in a northwesterly direction over the suburb of Clayton South.

Here is a brief description of the incident from Wikipedia:

> ... a class of students and a teacher from Westall High School (now Westall Secondary College) were just completing a sport activity on the main oval when an object, described as being a grey saucer-shaped craft with a slight purple hue and being about twice the size of a family car, was alleged to have been seen. Witness descriptions were mixed: Andrew Greenwood, a science teacher, told *The Dandenong Journal* at the time that he saw a silvery-green disc. According to witnesses the object was descending and then crossed and overflew the high school's south-west corner, going in a south-easterly direction, before disappearing from sight as it descended behind a stand of trees and into a paddock at The Grange in front of the Westall State School (primary students). After a short period (approximately 20 minutes) the object— with witnesses now numbering over 200—then climbed

241

at speed and departed towards the northwest. As the object gained altitude some accounts describe it as having been pursued from the scene by five unidentified aircraft which circled the object. Some described one disk, some claimed to have seen three.

An Australian newspaper article about the Westall High Scholl incident in 1966.

This is an astounding incident, one that was basically covered up by the authorities as quickly as possible. The children and teachers were asked not to speak very much about the incident, and Wikipedia says that witnesses and researchers were surprised when *The Sun News-Pictorial* (a tabloid) ran no story, yet *The Age* (a broadsheet, normal newspaper) did. Other newspapers ignored the story as well and *The Sun* and *The Herald* newspapers, while not mentioning the Westall incident, both published cartoons in the following day's editions that made light of the flying saucer phenomenon.

The one photo of the incident is a very good photo that appears to depict a large Haunebu flying saucer. This 1966 photo does not appear to be faked—and with an incident in broad daylight with over 200 witnesses, no one needs to fake a photo like this. It is interesting to note that two Australian UFO groups studied the incident intensely, as did the skeptic societies of Australia. None concluded that the incident was a hoax and the only suggestion that the skeptics were able to come up with was that it was an experimental military plane (sound familiar?) or a target dirigible being towed by a small plane.

So, what are we to make of this incident? Let us put aside any theories that a reptoid craft from Mars had stopped by to pick up some children for their underground slave labor factories. Rather we seem to have an incident that was meant to garner as much media attention as possible. It was a stunt—but not a hoax—of a power, the Third Power, to show off some of their aircraft to the world. It was a "hey, look at me" moment that involved landing a Haunebu on a sunny April morning near a high school in the very modern city of Melbourne, Australia.

It was staged so that the maximum number of people would witness the incident, adults and children, so that as much chatter as possible would be created and the media would have to take some notice and report it. Maybe photos of the event would even appear in newspapers and a larger section of the population would doubt what their government was telling them about flying saucers. But, sadly, the incident was largely suppressed outside of Australia, and even there major newspapers would not run the story. While many people in Western countries believe that our media is fair and free from government intrusion, this is sadly very wrong. It has been speculated that the *New York Times* ("All the news that's fit to print") has never run a story concerning UFOs.

Had this Haunebu made the flight from Antarctica to Melbourne? After its mission to shake up the Australian government and media did it then return to its base in Antarctica? Perhaps after that it made a mission to South America.

This Haunebu may have come from a hidden UFO bases in the Australian region. As I said earlier, it has been suggested that there are secret UFO bases on the island of New Guinea, for instance. This island is divided into two political divisions and is largely devoid of roads or towns. The interior of New Guinea could harbor a remote mountain base that would be served only by air. This would explain some of the UFO activity that has been reported on the island. The Solomon Islands are another possible site.

Since the Haunebu was capable of extended hours in flight it could reach any part of the world from whatever airbase it was stationed at. Whether that airbase was in Tibet, Antarctica, Greenland, Argentina or elsewhere, a craft taking off from that base would have been able to fly to any point on the globe, no matter how far. It would only be a matter of time for the craft to reach its destination. But we need to look here at why these craft would have been deployed to certain locations in the first place. What missions were they flying? Every flight must have had a purpose.

Chapter 7

A Saucer Called Vril

In hallways and in secret doorways
Where love's hiding places—with nowhere to go.
Goodbye to the fields and byways.
I remember saying I don't want to leave because
You were all there was to know about me.
—*Who Are You Now?*, Justin Hayward (Moody Blues)

We have talked about the Haunebu and the Vril craft. In the first chapter I discussed the curious name, Haunebu, and where it might have come from. The answer appears to be that the name originates in the term "Hauneburg Gerat" or "Hauneburg Device."

The origin of the name Vril is much more obvious, but what exactly was Vril and what was the mysterious Vril Society?

Hess, Haushofer and the Thule Society

Before the Vril Society was the earlier Thule Society, of which both Rudolf Hess and Karl Haushofer were members. Hess would go to Antarctica and then fly to Scotland in 1941 to convince the British to allow Hitler to have Europe while the Germans would guarantee Britain her colonial empire. Hess would later die in Spandau Prison in Berlin in 1987, its only prisoner. Haushofer would commit suicide with this wife in the months after the war.

Rudolf Hess was born in Alexandria, Egypt, on April 26, 1894, the son of a prosperous wholesaler and exporter. He came to Germany for the first time when he was fourteen. At the age of 20 he volunteered for the German Army at the outbreak of World War I in 1914. This was partly to escape the control of his domineering father who had refused to let him go to a university and instead wanted him to be part of the family business. Young Rudolf Hess had other ideas.

Hess was wounded twice during the war, and later became an airplane pilot. Hess was a large and powerful man, now a battle-hardened killer, and after the war he joined the Freikorps, a right-wing organization of ex-soldiers for hire. The Freikorps were involved in violently putting down Communist uprisings in Germany, often by having literal fistfights in the street.

Hess began attending the University of Munich where he studied political science. At the university he met Professor Karl Haushofer and joined the Thule Society, a secret society of sorts that espoused Nordic supremacy, mystical Germanic views of antiquity (such as a belief in Atlantis and Tibetan masters), and anti-Semitic views in the sense that the Jewish Old Testament was not the most important book in the world.

According to such authors as Nicholas Goodrick-Clarke[28] and Joscelyn Godwin[35] the Thule Society was originally a "German study group" headed by Walter Nauhaus, a wounded World War I veteran turned art student from Berlin who had become a keeper of pedigrees for the Germanenorden (or "Order of Teutons"), a secret society founded in 1911 and formally named in the following year.

Professor Karl Haushofer and Rudolf Hess, circa 1920.

Secret societies were booming in Germany during this period and Nauhaus moved to Munich in 1917 where his "Thule Society" was to be a cover name for the Munich branch of the Germanenorden, but a schism in the order caused events to develop differently. In 1918, Nauhaus was contacted in Munich by Rudolf von Sebottendorf who was an occultist and newly elected head of the Bavarian province of the schismatic offshoot known as the Germanenorden Walvater of the Holy Grail. The two men became associates in a recruitment campaign, and Sebottendorf adopted Nauhaus's Thule Society as a cover name for his Munich lodge of the Germanenorden Walvater at its formal dedication on August 18, 1918.

"Thule" was a land located by Greco-Roman geographers in the farthest north (often displayed as Iceland). The Latin term "Ultima Thule" is mentioned by the Roman poet Virgil in his pastoral poems called the *Georgics*. Thule may have been the original name for Scandinavia, although Virgil simply uses it as an expression for the edge of the known world to the north. The Thule Society identified Ultima Thule as a lost ancient landmass in the extreme north, near Greenland or Iceland, said by Nazi mystics to be the capital of ancient Hyperborea. Atlantis fitted into these myths as well and the Nazi mystics believed that Atlantis had existed in the Northern Sea, rather than in the Azores or the Caribbean.

Wikipedia says that Hitler biographer Ian Kershaw said that the organization's "membership list ...reads like a Who's Who of early Nazi sympathizers and leading figures in Munich," including Rudolf Hess, Alfred Rosenberg, Hans Frank, Julius Lehmann, Gottfried Feder, Dietrich Eckart, and Karl Harrer.

Joscelyn Godwin says in *Arktos*[35] that there was a similar group in France known as the Veilleurs (Vigilants) and that Hess may have met with them. Says Godwin:

> At the same time as Rudolf Hess was studying with Karl Haushofer, an Alsatian chemist called René Schwaller (1887-1961) was organizing some of his Theosophical friends in Paris into a group with the motto "Hierarchy, Fraternity, Liberty." Its first public appearance, in 1919,

was with a review, *L'Affranchi*, numbered so as to seem like a continuation of the Theosophical Society's magazine of prewar days. The articles, signed by pseudonyms, treated the themes of social and spiritual renewal in the context of internationalist and somewhat mystical politics. There was praise for Woodrow Wilson's League of Nations, and discreet allusions were made to a coming Messiah.

Within the Affranchis were two inner circles: one a "Centre Apostolique," Theosophical in nature; the other, formed in 1918, called the "Mystic Group Tala," a word that René Guénon translates as "the link." We know nothing of its activities, but can scarcely pass over the similarity of the name with Thule... in his first book, *Les Nombres* (1916), Schwaller had discussed just one symbol, besides the numbers and basic geometrical figures: the swastika, which he calls an accentuation of the cross within a circle, representing the archetypal formative movement of any body around it axis. There are other parallels between Schwaller's group, renamed in July 1919 "Les Veilleurs" (the Vigilants), and the Thule-descended parties organized by Hess and eventually headed by Hitler: their warrior mentality, their antijudaism, their uniform of dark shirts, riding breeches and boots (which Schwaller claimed to have designed), their messianism, and the tile of *Chef* given to their leaders.

...[Pierre Mariel] wrote that the young Rudolf Hess was a member of the Veilleurs. I cannot be sure whether to believed this or not; but it is worth considering. Hess, whose movement in 1919 are virtually uncharted, was certainly aware of what was going on in Paris. The difference of language would have presented no difficulty either to Schwaller or to the young Hess, raised in Alexandria and educated in Switzerland. It is thinkable that, on his return to Germany, Hess set himself to create, upon the foundations of the Thule Society, a veritable parody of the Veilleurs. Likewise it is not merely thinkable but definite that Schwaller's world of ideas intersected at many points with that of Thule: a circumstance that has troubled more than one admirer of this Hermetic master.[35]

Whether Hess somehow helped start the Thule Society or not, Professor Karl Haushofer was certainly central to the secret society. He was a former general whose theories on expansionism and race formed the basis of the concept of Lebensraum—increased living space for Germans at the expense of other nations. He advocated extending Greater Germany to Moscow and the oilfields of Baku. Haushofer's teachings were very influential in Germany's ultimate invasion of Eastern Europe under the belief that the Germans

The logo of the Thule Society with oak leaves, dagger and swastika.

Professor Karl Haushofer.

needed to expand their territory and culture. Their allies in Japan had a similar belief in Japan's alarming Imperial expansion throughout eastern Asia and the Western Pacific.

Haushofer was well known to the German public. He was a prolific writer, publishing hundreds of articles, reviews, commentaries, obituaries and books, many of which were on Asian topics. He claimed to have been to Tibet and to have been initiated in secret Tibetan rites. If he had been to Tibet, he might have been instrumental in setting the SS airbase there. Perhaps he even flew on a Haunebu to the Tibetan SS base, as he was apparently very fond of the country. Haushofer also arranged for many leaders in the Nazi Party and in the German military to receive copies of his various works.

Louis Pauwels, in his book *Monsieur Gurdjieff*, describes Haushofer as a former student of George Gurdjieff. Gurdjieff also claimed to have been initiated in Tibet. Pauwels and others have said that Haushofer created the Vril Society from the Thule Society.

The Nazi regime began to persecute Jews soon after its seizure of power. Hess's office was partly responsible for drafting Hitler's Nuremberg Laws of 1935, laws that had far-reaching implications for the Jews of Germany, banning marriage between non-Jewish and Jewish Germans and depriving non-Aryans of their German citizenship. Haushofer remained friendly with Hess, who protected Haushofer and his wife, who was part Jewish, from the strict racial laws implemented by the Nazis, which deemed her a "half-Jew." Hess issued documents exempting them from the new legislation.

During the prewar years, Haushofer was instrumental in linking Japan to the Axis powers, acting in accordance with the theories of his book *Geopolitics of the Pacific Ocean*. Haushofer apparently made several trips to Japan prior to 1941 and is said to have been part of a Japanese secret society known as the Black Dragon Society. Several American films and serials were made just prior to WWII and during the war that featured this alleged secret society. In some ways the Black Dragon Society was a Japanese version of the Vril Society (which Haushofer allegedly founded).

Along these lines, it is interesting to surmise that some of the UFO activity seen and photographed in Japan starting in the 1940s might have to do with Haunebu or Vril craft picking up members of the Black Dragon Society in the decades after the war. It may well be that Miguel Serrano was also a member of the Black Dragon Society and may have taken a Haunebu flight to Japan from India or even Chile.

In 1944, after the July 20 Plot to assassinate Hitler failed, Haushofer's son Albrecht (1903–1945), a close friend of Rudolf Hess, was implicated, in part because of his association with Hess. Albrecht went into hiding but was arrested on December 7, 1944

Hess, Heinrich Himmler, Phillip Bouhler, Fritz Todt, Reinhard Heydrich, and others listening to Konrad Meyer at a Generalplan Ost exhibition, March 20, 1941.

251

and put into the Moabit prison in Berlin. On the night of 22–23 April 1945, as Soviet troops were already reaching Berlin's eastern outskirts, he and other prisoners who were deemed part of the July 20 Plot, such as Klaus Bonhoeffer and Rüdiger Schleicher, were walked out of the prison with other prisoners by an SS-squad and killed with a gunshot wound to the neck. The only eyewitness to these murders was another political prisoner, Herbert Kosney, who managed to move his head at the last moment so that the shot meant for his neck missed.

After the war Karl Haushofer continued to live at his Hartschimmelhof estate at Pähl/Ammersee in occupied West Germany. Starting on September 24, 1945, some four months after the war had ended, Haushofer was informally interrogated by Father Edmund A. Walsh on behalf of the Allied forces to determine whether he should stand trial for war crimes. After some time, Father Walsh determined that he had not committed any.

On the night of 10–11 March 1946, Karl Haushofer and his wife committed suicide in a secluded hollow on their estate. Both drank arsenic and his wife then hanged herself. In the book *The Morning of the Magicians*.[29] the authors claimed that Haushofer committed Japanese-style *hari-kari* suicide with a Japanese sword but Joscelyn Godwin in *Arktos*[35] says that this is one of a number of questionable things in that book.

Haushofer's influence on Rudolf Hess was considerable. Haushofer had befriended Hess and influenced him to join (or co-create) the Thule Society circa 1918. Then, on July 1, 1920, Hess heard Adolf Hitler speak in a small Munich beer hall, and immediately joined the Nazi Party, becoming the sixteenth member. After his first meeting with Hitler, Hess said he felt "as though overcome by a vision." Hess was to become utterly devoted to Hitler and eagerly agreed with everything the shouting politician said. At early Nazi Party rallies, Hess was a formidable fighter who continually brawled with Marxist activists and others who violently attempted to disrupt Hitler's speeches.

In 1923, Hess took part in Hitler's failed Beer Hall Putsch. Hitler and the Nazis attempted to seize control of the government of Bavaria at this time and Hess was arrested and imprisoned along with Hitler at Landsberg Prison. While the two were in prison, Hess took dictation for Hitler's book, *Mein Kampf.* Hess also made some editorial suggestions regarding the organization of the Nazi Party, the notion of Lebensraum, plus material in the book about the historical role of the British Empire.

Both Hitler and Hess were released from prison in 1925 and Hess served for several years as the personal secretary to Hitler in spite of having no official rank in the

The emblem of the expedition.

Nazi Party. In 1932, Hitler appointed Hess an SS General and the Chairman of the Central Political Commission of the Nazi Party as a reward for his loyal service. On April 21, 1933, Hess was made Deputy Führer, a figurehead position with mostly ceremonial duties.

Hess was motivated by his loyalty to Hitler and a desire to be useful to him; he did not seek power or prestige or take advantage of his position to accumulate personal wealth. He and his wife Ilse lived in a modest house in Munich. Although Hess had less influence than other top Nazi officials, he was popular with the masses.

Hess then took part in the 1939 secret German expedition to Antarctica. In December of 1938 Hermann Göring launched an expedition to Antarctica in an effort to establish a naval base

A seaplane launches off of the German ship *Schwabenland*.

A group from the expedition planting the Nazi flag in Neuschwabenland.

and whaling station on the polar continent. The New Swabia Expedition left Hamburg for Antarctica aboard MS *Schwabenland* on December 17, 1938. The MS *Schwabenland* was a freighter built in 1925 and renamed in 1934 after the Swabia region in southern Germany. The MS *Schwabenland* was also able to carry special aircraft that could be catapulted from the deck.

The expedition was top secret and was overseen by Göring himself. The Thule Society was also apparently involved. The expedition had 33 members plus the *Schwabenland*'s crew of 24. On January 19, 1939 the ship arrived at the Princess Martha Coast of Antarctica, in an area which had recently been claimed by Norway as Queen Maud Land, and began charting the region. Nazi German flags were placed on the sea ice along the coast. Naming the area Neuschwabenland after the ship, the expedition established a temporary base, and in the following weeks teams walked along the coast recording claim reservations on hills and other significant landmarks.

Upon his return in April of 1939 Hess reported to Göring and Hitler what had been discovered in Antarctica and whether they had found a suitable place for a German naval base. Such a base would serve commercial ships such as whalers as well as military ships and submarines.

This success must have one of the reasons that Hitler made

Hess second in line to succeed him, after Hermann Göring, in September of 1939. During this same time, Hitler appointed Hess's chief of staff, Martin Bormann, as his personal secretary, a post formerly held by Hess. The German invasion of Poland happened at this time as well.

That Antarctica was part of Göring, Hess and Haushofer's plan for German expansion and the recreation of a global German colonial community is quite clear. Joseph Farrell quotes the German author Heinz Schön, who wrote a 2004 book in German called *Mythos Neuschwabenland: Für Hitler am Südpol: Die deutsche Antarktis-expedition 1938/39* and says:

> As commissioner for the Four-Year Plan, Göring knew the importance for Germany of whaling in Antarctica, and how essential it was to ensure this, and to open up new fishing grounds. It seemed high time for him to send a large expedition to Antarctica. On May 9, 1938, a plan for an Antarctic expedition, drawn up by the staff of his ministry, which was to be carried out in the Antarctic summer of 1938/39, was presented to him. He approved, and commissioned Helmut Wohlthat as Minister-Director for special projects, with the preparation of the expedition, and conferred upon him all his powers of authority.[2]

As noted above, one of the main purposes of the secret expedition was to find an area in Antarctica for a German whaling station, as a way to increase Germany's production of fat. Whale oil was then the most important raw material for the production of margarine and soap in Germany and the country was the second largest purchaser of Norwegian whale oil, importing some 200,000 metric tons annually.

Germany did not want to be dependent on imports and it was thought that Germany would soon be at war, which would put a lot of strain on Germany's foreign currency reserves. The other goal of this secret expedition was to scout possible locations for a German naval base, and that would include a base for submarines.

Hess in Scotland and Aktion Hess

It is said that Hess was obsessed with his health to the point of hypochondria, consulting many doctors and other practitioners for what he described to his British captors as a long list of ailments involving the kidneys, colon, gall bladder, bowels and his heart. Hess was a vegetarian, like Hitler and Himmler, and he did not smoke or drink. He was a big believer in homeopathic medicines and Rudolf Steiner-type food that was "biologically dynamic." Hess was interested in music, enjoyed reading and loved to spend time hiking and climbing in the mountains with his wife, Ilse. He and his friend Albrecht Haushofer, Karl Haushofer's son, shared an interest in astrology, psychic powers, clairvoyance and the occult.

As the war progressed, Hitler's attention became focused on foreign affairs and the conduct of the war. Hess, who was not directly engaged in these endeavors, became increasingly sidelined from the affairs of the nation and from Hitler's attention. Martin Bormann had successfully supplanted Hess in many of his duties and essentially usurped the position at Hitler's side that Hess had once held. Hess later said that he was concerned that Germany would face a war on two fronts as plans progressed for Operation Barbarossa, the invasion of the Soviet Union scheduled to take place in 1941.

Hess decided to attempt to bring Britain to the negotiating table by travelling there himself to seek direct meetings with the members of the British government. He asked the advice of Albrecht Haushofer who suggested several potential contacts in Britain. Hess settled on fellow aviator Douglas Douglas-Hamilton, the Duke of Hamilton, whom he had met briefly during the Berlin Olympics in 1936.

On Hess's instructions, Haushofer wrote to Hamilton in September of 1940, but the letter was intercepted by MI5 and Hamilton did not see it until March of 1941. The Duke of Hamilton was chosen because he was one of the leaders of an opposition party that was opposed to war with Germany, and because he was a friend of Karl Haushofer. In a letter that Hess wrote to his wife dated November 4, 1940, he says that in spite of not receiving a reply from Hamilton, he intended to proceed with his plan to fly himself to Scotland and meet with Hamilton.

Hess began training on the Messerschmitt 110, a two-seater twin-engine aircraft, in October of 1940 under the chief test pilot at Messerschmitt. He continued to practice, including logging many cross-country flights, and found a specific aircraft that handled well—a Bf 110E-1/N—which was from then on held in reserve for his personal use. He asked for a radio compass, modifications to the oxygen delivery system, and large long-range fuel tanks to be installed on this plane, and these requests were granted in March 1941.

Hoping to save the Reich from disaster and redeem himself in the eyes of his Führer, Hess put on a Luftwaffe uniform and leather jacket (he was an SS General as well) and flew the fighter plane alone toward Scotland on a 'peace' mission on May 10, 1941, just before the Nazi invasion of the Soviet Union. After a final check of the weather reports for Germany and the North Sea, Hess took off at 17:45 from the airfield at Augsburg-Haunstetten in his specially prepared aircraft.

It was the last of several attempts by Hess to fly to Scotland, with previous efforts having been called off due to mechanical problems or poor weather. Wearing the leather flying suit, he brought along a supply of money and toiletries, a flashlight, a camera, maps and charts, and a collection of 28 different medicines, as well as dextrose tablets to help ward off fatigue, and an assortment of homeopathic remedies.

Hess flew north and when he reached the west coast of Germany near the Frisian Islands, he turned and flew in an easterly direction for twenty minutes to stay out of range of British radar. He then took a heading of 335 degrees for the trip across the North Sea, initially at low altitude, but travelling for most of the journey at 5,000 feet (1,500 meters). At 20:58 he changed his heading to 245 degrees, intending to approach the coast of northeast England near the town of Bamburgh, Northumberland.

Flying north after reaching the west coast of Germany, Hess approached the coast of northeast England near the town of Bamburgh and realized that sunset was still nearly an hour away, and he needed darkness to fly past the coast. Hess backtracked, zigzagging back and forth for 40 minutes until it grew dark. Around this time his auxiliary fuel tanks were exhausted, so he released them into the sea. Shortly after that he was over Scotland and at 6,000 feet Hess bailed out and parachuted safely to the ground where he encountered a Scottish farmer and told him in English, "I have an important message for the Duke of Hamilton."

Now in captivity, Hess told his captors that he wanted to convince the British government that Hitler only wanted Lebensraum for the German people and had no wish to destroy a fellow 'Nordic' nation. He also knew of Hitler's plans to attack the Soviet Union and wanted to prevent Germany from getting involved in a two-front war, fighting the Soviets to the east and Britain and its allies in the west.

During interrogation in a British Army barracks, Hess proposed that if the British would allow Nazi Germany to dominate Europe, then the British Empire would not be further molested by Hitler. Hess demanded a free hand for Germany in Europe and the return of former German colonies as compensation for Germany's promise to respect the integrity of the British Empire. Hess insisted that German victory was inevitable and said that the British people would be starved to death by a Nazi blockade around the British Isles unless they accepted his generous peace offer.

Before his departure from Germany, Hess had given his adjutant, Karlheinz Pintsch, a letter addressed to Hitler that detailed his intentions to open peace negotiations with the British. Pintsch delivered the letter to Hitler at the Berghof (Hitler's home in the Bavarian Alps where he spent a great deal of time during WWII and which became an important center of government) around noon on May 11. After reading the letter, Hitler let loose an angry yell that was heard throughout the entire Berghof, and sent for a number of his inner circle as he was concerned that a putsch (an attempt to overthrow the government) might be underway.

Hess's odd flight out of Germany, but not his destination or fate, was first announced by Munich Radio in Germany on the evening of May 12. On May 13 Hitler sent Foreign Minister Joachim von Ribbentrop to give the news in person to Mussolini, and on the same day the British press was permitted to release full information about the events. Ilse Hess finally learned that her husband had survived the trip when news of his fate was broadcast on German radio on May 14. Hitler publicly accused Hess of suffering from "pacifist delusions."

Hitler was worried that his allies, Italy and Japan, would perceive Hess's act as an attempt by Hitler to secretly open peace negotiations with the British. Hitler contacted Mussolini specifically to reassure him otherwise. Hitler ordered that the German press should characterize Hess as a madman who made the decision to fly to Scotland entirely on his own, without Hitler's knowledge. Subsequent German newspaper reports described Hess as "deluded, deranged," indicating that his mental health had been affected by injuries sustained during World War I.

Some members of the government, such as Göring and Propaganda Minister Joseph Goebbels, believed this only made matters worse, because if Hess truly were mentally ill, he should not have been holding such an important government position, second in line to succeed the Führer.

Hitler stripped Hess of all of his party and state offices, and secretly ordered him shot on sight if he ever returned to Germany. He abolished the post of Deputy Führer, assigning Hess's former duties to Bormann, with the title of Head of the Party Chancellery. Bormann used the opportunity afforded by Hess's departure to secure significant power for himself. Meanwhile, Hitler initiated Aktion Hess, a flurry of hundreds of arrests of astrologers, faith healers and occultists that took place around June 9 and 10. The campaign was part of a propaganda effort by Goebbels and others to denigrate Hess and to make scapegoats of occult practitioners.

This process involved rounding up and imprisoning Hess's associates, including his wide-ranging network of occultists, astrologers, and ritualists. By positioning himself squarely at the center of the occult movement and then falling from grace so spectacularly, Hess doomed his fellow practitioners to a very sudden end. Everything from fortune-telling to astrology was outlawed, and the Nazi party's infatuation with the occult was over. However, the occult undercurrent remained in the SS, of which Hess had been a General.

One man who believed in Hitler's almost mythical status was a college professor named Johann Dietrich Eckhart. He was a member of the mysterious Thule Society, and he and many of the group's members believed that a German messiah was prophesied to enter history in the near future. This German leader would return the nation to its former glory, and avenge its defeat in the First World War, undoing the humiliation imposed upon the country with the Treaty of Versailles. Eckhart was a student of eastern mysticism and developed an ideology of a "genius superman," based on writings by the Völkisch author Jörg Lanz von Liebenfels. Eckhart met Hitler in 1919 and was certain that this man was the savior he believed Germany had been promised. The man went on

260   Johann Dietrich Eckhart.

to shape Hitler's ideologies considerably, sculpting the beliefs and worldview of the Nazi party. Eckhart died in 1923 before Hitler and the Nazis could come to power.

Hitler has often been seen as the subject of various prophecies and the Nazis had sought to use the prophecies of Nostradamus for their own benefit. Perhaps most notable is the Nostradamus verse in which he wrote, "Beasts ferocious with hunger will cross the rivers; the greater part of the battlefield will be against Hister," along with references to a "Child of Germany."

One person arrested during Aktion Hess was Karl Ernst Krafft, an astrologer and psychic who claimed he was clairvoyant. He was a committed supporter of the Nazi regime, but in 1939 he made a prediction of an assassination attempt against Adolf Hitler between the 7th and 10th of November of that year.

At the time his claims received little attention, but following the detonation of a bomb in the Munich Beer Hall on November 8, everything changed. Hitler had already left the building by the time the explosion occurred. It killed seven people and injured almost 70 more, but the target of the attack escaped unscathed. Soon afterward, word of Krafft's prophecy reached Rudolf Hess, and the fortune-teller was arrested. However, he managed to convince his interrogators that he was innocent of any wrongdoing and that his gifts of prophecy were genuine.

Krafft was well-liked by Hitler himself, and was ordered to begin an evaluation of the prophecies of Nostradamus that would favor the Nazi worldview. However, his own gifts were his undoing; as mentioned above, following Rudolf Hess's flight to Scotland, he was swept up in Aktion Hess. Krafft was arrested in 1941 and died in prison in 1945.

Vril Power and the Vril Staff

One has to wonder if Aktion Hess made the Vril Society become more secretive. The Thule Society had become inactive before Aktion Hess, barely surviving into the 1930s, and now it was the Vril Society that commanded the ear of powerful SS Generals such as Heinrich Himmler and Martin Bormann. All of these men were familiar with the concept of "vril." Vril was an energy and there was also a "Vril Staff."

261

The concept of an energy called vril comes from an 1871 novel titled *The Coming Race* by Edward Bulwer-Lytton, published anonymously in Britain. It has also been published as *Vril, the Power of the Coming Race*.[53] Vril was a fabled source of free and infinite energy, a sort of electricity all around us—similar to ether—that could be utilized. Something like static electricity but much more powerful and useful. The term vril was first used in the 1871 novel by Edward Bulwer-Lytton, *Vril: The Coming Race*.[53] The book is about an inner-earth dwelling race of superhumans called the Vril-ya, who master this vril energy for its healing, positive and destructive properties.

The novel centers on the narrator, who is an unnamed young, wealthy traveller who visits a friend who is a mining engineer. They explore a natural chasm in a mine that has been exposed by an exploratory shaft using a rope. The narrator reaches the bottom of the chasm safely, but the rope breaks and the mining engineer is killed. The narrator then finds his way into a subterranean world occupied by beings who seem to be advanced humans. He befriends the first being he meets, and this person guides him around a city that has features similar to ancient Egyptian architecture. The explorer meets his host's wife, two sons and daughter who learn to speak English by way of a makeshift dictionary. His guide comes toward him, and he and his daughter, Zee, explain who they are and how they function.

The narrator discovers that these beings, who call themselves Vril-ya, have great telepathic and other parapsychological abilities, such as being able to transmit information, get rid of pain, and put others to sleep.

The narrator soon discovers that the Vril-ya are descendants of an antediluvian civilization called the Ana, and live in networks of caverns that are linked by tunnels. Originally surface dwellers, they fled underground thousands of years ago to escape a massive flood and gained greater power by facing and dominating the harsh conditions of the Earth. The place where the narrator descended houses 12,000 families, one of the largest groups. Their society uses vril and those that are spiritually elevated are able to master this power—an extraordinary force that can be controlled at will.

The uses of vril in the novel amongst the Vril-ya vary from

destruction to healing. The Vril Staff can be used in focusing vril energy. According to Zee, the daughter of the narrator's host, vril can be changed into the mightiest agency over all types of matter, both animate and inanimate. It can destroy like lightning or replenish life, heal, or cure. It is used to rend ways through solid matter. Its light is said to be steadier, softer and healthier than that from any flammable material. It can also be used as a power source for animating mechanisms. Vril can be harnessed by use of the Vril Staff or mental concentration. In many ways it is similar to

Edward George Earle Bulwer-Lytton.

"the Force" talked about by Jedi Knights in the *Star Wars* movies.

The caduceus—like a Vril Staff?

The book describes a Vril Staff as an object in the shape of a wand or a staff that is used as a channel for vril. The narrator describes it as hollow with "stops," "keys," or "springs" in which vril can be altered, modified, or directed to either destroy or heal. The staff is about the size of a walking stick but can be lengthened or shortened according to the user's preferences. The appearance and function of the Vril Staff differs according to gender, age, etc. Some staves are more potent for destruction; others, for healing. The staves of children are said to be much simpler than those of sages; in those of wives and mothers, the destructive part is removed while the healing aspects are emphasized.

It seems that the Vril Staff is similar to the Egyptian staves and pillars such as the *Was* staff or the *Djed* pillar. An early version of this is also seen in Sumeria. In Bolivia, the figure of Virococha on the Sun Gate at Tiwanaku is holding two staffs on either side of his body and the first Inca, Manco Capac was said to carry a staff; he sunk it into the ground at Cuzco when he arrived at that place and declared it the capital of the new Inca Empire. We might also envision this staff as functioning in a similar manner to a caduceus, the symbol of a staff with two snakes intertwined around it and sometimes with wings at the top (the medical staff symbol—the Rod of Asclepius—has only one snake on it). The caduceus is the staff carried by Hermes in Greek mythology and consequently by Hermes Trismegistus in Greco-Egyptian mythology. It is also depicted being carried in the left hand of Mercury, the Roman messenger of the gods.

Vril, the Power of the Coming Race was an important and popular book when it was published, and when it was learned that the anonymous author was actually the well-liked royal figure,

Lord Edward Bulwer-Lytton, the book became even more popular. Other occultists of the time immediately recognized the imagery and concepts in the book as esoteric Atlantean science thinly veiled in a novel. These occultists, including Helena Blavatsky, accepted the book as based on occult truth, but did not believe in the civilization living in caverns beneath the planet (though they did believe in an underground world) as much as they did in ancient civilizations and the use of some electric force that was also able lift huge blocks of stone that weighed hundreds of tons each. Surely, only a force like vril could lift such massive blocks of granite with a force of anti-gravity that could magically float these blocks into place.

We see such gigantic blocks of stone all over the world and they are always very ancient. Such places as Baalbek, Egypt, Carnac, Stonehenge, Teotihuacan, Sacsayhuaman, Tiwanaku, Easter Island contain megalithic structures or statues that defy the logic of a modern engineer and make the idea of levitation a reality. Is vril power the ability to control gravity and the space around us?

The book and the concept of vril inspired the creation of the Vril Society out of the Thule Society. It was taught that vril was a "Cosmic Primal Force"—an exotic spiritual technology that would bring about a new Utopian era for humanity: the Rebirth of Atlantis. But little is known of the Vril Society and some historians doubt that it ever existed.

Godwin says that Willy Ley, a German rocket engineer who emigrated to the United States in 1937, published an article titled "Pseudoscience in Naziland" in 1947 in the magazine *Astounding Science Fiction*. He wrote that the high popularity of irrational convictions in Germany at that time explained how National Socialism could have fallen on such fertile ground. Among various pseudoscientific groups he mentions

Astounding Science Fiction, Aug. 1947.

265

is one that looked for the power of vril:

> The next group was literally founded upon a novel. That group which I think called itself 'Wahrheitsgesellschaft'—Society for Truth—and which was more or less localized in Berlin, devoted its spare time looking for Vril. Yes, their convictions were founded upon Bulwer-Lytton's "The Coming Race." They knew the book was fiction, Bulwer-Lytton had used that device in order to be able to tell the truth about this "power." The subterranean humanity was nonsense, Vril was not. Possibly it had enabled the British, who kept it as a State secret, to amass their colonial empire. Surely the Romans had had it, enclosed in small metal balls, which guarded their homes and were referred to as *lares*. For reasons which I failed to penetrate, the secret of Vril could be found by contemplating the structure of an apple, sliced in halves.
>
> No I am not joking, that is what I was told with great solemnity and secrecy. Such a group actually existed, they even got out the first issue of a magazine which was to proclaim their credo.[35]

As far as the vril apple meditation, Godwin mentions that Rudolf Steiner suggests such a meditation in his books, which is to slice an apple in half across the middle which would reveal a five-pointed star. If an apple is sliced down the center starting from the stem, down through the middle, it reveals what some might visualize as a vortex-toroid. Either could be part of some "vril" meditation.

The Vril Society was discussed in 1960 by Jacques Bergier and Louis Pauwels in their book *The Morning of the Magicians*[29] originally published in French. In the book they claimed that the Vril Society was a secret community of occultists in pre-Nazi Berlin that was a sort of inner circle of the Thule Society, of which Rudolf Hess was a member. They also thought that it was in close contact with the English group known as the Hermetic Order of the Golden Dawn. They also mention the Tibetans living in Berlin as discussed earlier. They do not mention a Nazi base in Tibet or

Tibetans at any Arctic base as per Landig.

In a separate book, only published in French, *Monsieur Gurdjieff*, Louis Pauwels claims that a Vril Society was "founded by General Karl Haushofer, a student of Russian magician and metaphysician Georges Gurdjieff."

In *Black Sun*,[28] Nicholas Goodrick-Clarke refers to the research of the German author Peter Bahn. Bahn writes in his 1996 essay, "Das Geheimnis der Vril-Energie" ("The Secret of Vril Energy"), of his discovery of an obscure esoteric group calling itself the Reichsarbeitsgemeinschaft Das Kommende Deutschland (Imperial Working Society of the Coming Germany), or RAG for short. RAG taught its readers to meditate on the image of an apple sliced vertically in half, representing the map of universal free energy available from Earth's magnetic field. Using this knowledge they

Vermutlich Schauberger-Gerät mit Winter-Tarnanstrich
(Raum Augsburg 1939)

The caption says that this Schauberger Device was flown in 1939.

This photo of a Vril was supposedly taken in 1934.

would harness vril, the "all-force of the forces of nature" by using "ball-shaped power generators" to channel the "constant flow of free radiant energy between outer space and Earth." This was just as Willy Ley recalled.

RAG published several small booklets on vril energy. One of the booklets was published in 1930 and entitled *Vril, Die Kosmische Urkraft* (Vril, the Cosmic Elementary Power) written by a member of this Berlin-based group, under the pseudonym "Johannes Täufer" which in German translates to "John [the] Baptist."

The booklets describe Atlantean dynamo-technology superior to modern mechanistic science, saying that their spiritual technology and limitless vril energy is what enabled Egyptians and Mayans to build massive pyramids. Nikola Tesla spoke of similar things in his interviews with reporters at his apartment in New York. All manners of electrical power and uses were being examined by the Nazis, including "vril."

The booklets on vril were issued by the influential German astrological publisher by the name of Otto Wilhelm Barth (whom Bahn believes was "Täufer"). The 60-page pamphlet says little of the group other than that it was founded in 1925 to study the uses of vril energy.

So we see how RAG was really what we call "the Vril Society" and that it was a real organization. They became known as the

Vril Society because of the early Willy Ley article in 1947. It was the publication of the notable booklet on vril that caught people's notice at the time, in professor Ley, who left Germany in 1937 like many academics. The 1925 date is interesting as that date is often given as to the vague when the Thule Society stopped its meetings. It seems that the inner core of the Thule Society became RAG and was nicknamed the Vril Society. They apparently operated from 1925 until the end of the war in 1945.

At this time, with Germany being overrun by the Allies from the east and west, this society apparently dissolved and some members literally flew off in flying saucers to secret SS bases around the world (Karl Haushofer committed suicide). The SS airbase in Tibet would have been a safe place in 1945 when Tibet was still an independent country. Any relationship that the SS had with Tibet would have continued as it was—even after the war— at least for a few years and probably into the 1950s. China invaded Tibet—a de facto country allied with Germany during the war—in October of 1950, but it took years to secure even the most basic parts of the country, such as border areas and major towns that are stretched out over vast unpopulated areas, similar to Patagonia.

So, we see that the Vril Society was a real secret society that published at least two booklets on "vril power." When one

A photo of a Vril craft with a car in the background.

A photo of a Haunebu taken through a windshield.

adds the work of Nikola Tesla and the many designs for electric aircraft, including descriptions by Tesla, one can see that the war machine of the Nazis could use some technology like this if it were shown to be practical. Other scientists in Germany, Austria and elsewhere were working on exotic technologies as has been alleged of Guglielmo Marconi and his associates. Death rays were popular during the 1920s and 1930s and the Nazis were aware of these technological studies and theories, and wanted to pursue them. Allegedly the Americans were working on the so-called Philadelphia Experiment, which is beyond the scope of this book.

Once elements of the SS were interested in such devices as "the Bell" and electric flying saucers, funding for these programs would be delivered, and prototypes—at great expense—were created. It seems to have been in the 1930s that money was invested in the creation of the Haunebu and Vril craft, as well as in other experiments such as the Bell. Meanwhile, the development of V-1 rockets, long-range bombers and the atomic bomb proceeded on a separate track.

That a Haunebu craft was already being tested in the 1930s can be seen in a YouTube video of a 1939 flight test. The video can be seen at https://youtu.be/jywwCaph25c (or do a search for Haunebu Test Flight) and it is an amazing piece of film. The

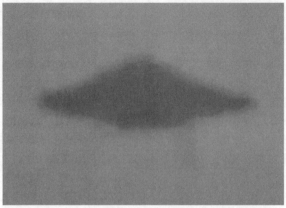

A screenshot of the YouTube video showing a Haunebu fly.

YouTube video is one minute and ten seconds in length but the footage of the Haunebu is only about 13 seconds long. The rest of the video shows it over and over again, zoomed and slowed down. The video shows a Haunebu lifting off in an area where there are hangars and other buildings, hovering in the air for a few seconds, and then landing. A banner beneath the video says: HAUNEBU Test Flight, NAZI Germany, 1939.

The footage looks authentic and it adds credence to the early creation of the Haunebu and Vril craft before the war. As long as we can agree that many UFO photos and descriptions are valid and that something is clearly going on in this regard, then we see how the stories, including photos, diagrams and film footage, of

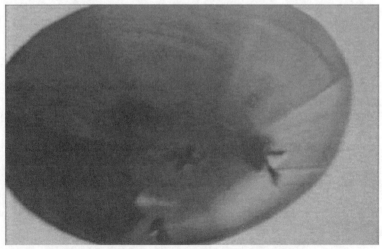

A photo of the underside a Vril craft in flight.

271

German flying saucers comprise a credible subject. The Vril Society is also a credible subject, whatever its activities were.

Some stories of the Vril Society may not be so credible however, such as the tales of the Vril Maidens. Nicholas Goodrick-Clarke tells us in his scholarly book *Black Sun*[28] that the story of the Vril Maidens supposedly begins in 1919 when an inner group of Thule and Vril Society members held a meeting in Vienna, where they met with a psychic named Maria Orsic. She allegedly presented transcripts of automatic writing she received in a language that she didn't understand. The language was found to be ancient Sumerian, allegedly channeled from a planet in the Aldebaran solar system—Aldebaran being the brightest star in the constellation of Taurus—68 light years away from Earth.

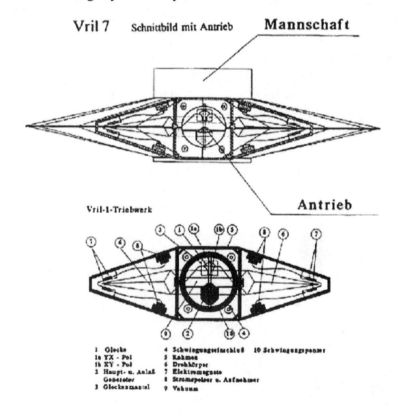

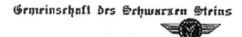

The plans for the Vril 7 craft.

The Vril Maidens—an inner circle of young female psychic mediums, led by the beautiful Maria Orsic—kept their hair long because they believed it to be an extension of the nervous system which acted as an antenna when telepathically communicating between worlds.

The Vril Maidens

The Vril Maidens were said to channel blueprints for time travel machines, anti-gravity technology, and more. Over the decades this supposedly led to the development of the Vril and Haunebu series of anti-gravity flying saucers.[28]

According to information on a "vril site" that promotes Maria Orsic on the Internet, the Vril Society was formed by a group of female psychic mediums led by the Thule Gesellschaft medium Maria Orsitsch (Orsic) of Zagreb. She claimed to have received communication from Aryan aliens living on Alpha Centauri, in the Aldebaran system. Allegedly, these aliens had visited Earth and settled in Sumeria, and the word "vril" was formed from the ancient Sumerian word "Vri-Il" ("like god"). Says the text:

> A second medium was known only as Sigrun, a name etymologically related to Sigrune, a Valkyrie and one of Wotan's nine daughters in Norse legend. The Society allegedly taught concentration exercises designed to awaken the forces of Vril, and their main goal was to achieve Raumflug (Spaceflight) to reach Aldebaran. To achieve this, the Vril Society joined the Thule Gesellschaft to fund an ambitious program involving an inter-dimensional flight machine based on psychic revelations from the Aldebaran aliens. Members of the Vril Society are said to have included Adolf Hitler, Alfred Rosenberg, Heinrich Himmler, Hermann Göring, and Hitler's personal physician, Dr. Theodor Morell. These were original members of the Thule Society which supposedly joined Vril in 1919. The NSDAP (NationalSozialistische Deutsche ArbeiterPartei) was created by Thule in 1920, one year later. Dr. Krohn, who helped to create the Nazi flag, was also a Thulist. With Hitler in power in 1933, both

An alleged photo of the pyschic Maria Orsic.

Thule and Vril Gesellschafts allegedly received official state backing for continued disc development programs aimed at both spaceflight and possibly a war machine.

After 1941 Hitler forbade secret societies, so both Thule and Vril were documented under the SS E-IV unit. The claim of an ability to travel in some inter-dimensional mode is similar to Vril claims of channeled flight with the Jenseitsflugmaschine (Other World Flight Machine) and the Vril Flugscheiben (Flight Discs).

It is difficult to verify this information on the Vril Society or Maria Orsic. Some believe that she was born in Zagreb on October 31, 1894 and visited Rudolf Hess in Munich in 1924. She is said to have disappeared in 1945. An Internet biography of her states:

In December 1943 Maria attended, together with Sigrun, [another psychic, even more mysterious than Orsic] a meeting held by Vril at the seaside resort of Kolberg. The main purpose of the meeting was to deal with the Aldebaran project. The Vril mediums had received precise information regarding the habitable planets around the sun Aldebaran and they were willing to plan a trip there. This project was discussed again the 22nd January 1944 in a meeting between Hitler, Himmler, Dr. W. Schumann (scientist and professor in the Technical University of Munich) and Kunkel of the Vril Gesellschaft. It was decided that a *Vril 7 Jaeger* would be sent through a dimension channel independent of the speed of light to Aldebaran. According to N. Ratthofer (writer), a first test flight in the dimension channel took place in late 1944.

Maria Orsic disappeared in 1945. The 11th of March of 1945 an internal document of the Vril Gesellschaft was sent to all its members; a letter written by Maria Orsic. The letter ends: "niemand bleibt hier" [no one stays here]. This was the last announcement from Vril. It is speculated they departed to Aldebaran.

It seems that the Vril Society was real but the stories about

275

Maria Orsic and the contact with the planet Aldebaran are doubted by many including Nicholas Goodrick-Clarke.[28] Some researchers claim that nothing can be found about Maria Orsic before 1990. However, it does seem that one of the saucer craft built by the Germans was given the name Vril. Henry Stevens says:

> The late Heiner Gehring wrote to me that he had been told that the efforts of the Vril Society at channeling continued after the war, until 1946. It is no secret that in the lore propounded by Ralf Ettl and Norbert Juergen-Ratthofer elements of the Nazi regime allegedly associated with the Vril Society launched an interdimensional flight to an extraterrestrial world seeking help in the Second World War. This flight communicated, allegedly, via medial transmission (or channeling). This medial contact, according to Heiner's source, continued until 1946 with the living crew of that ship when it was abruptly terminated. Of course, there is no way to verify any of this.[46]

More likely, Vril craft were not sent through some interdimensional doorway to Aldebaran, nor did members of the Vril Society probably escape to this planet. Far more likely, if members of the Vril Society had something to do with the actual manufacture and operation of these flying saucers, they would

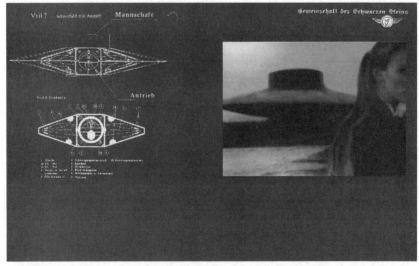

The plans for the Vril 7 craft with Maria Orsic superimposed on the right.

have escaped to various SS bases still in operation at the end of the war and afterward. Since the SS was clearly in control of these craft, plus some long-range aircraft and submarines, any Vril Society members wanting to escape from Germany at the end of the war would have had to hitch a ride on one of these SS special craft. It is likely that most Vril Society members in Germany and Austria at the end of the war remained where they were, like Professor Karl Haushofer, allegedly one of the founding members of the Vril Society.

What seems to be powering the Haunebu and the Vril is some sort of plasma (electrified gas such as mercury) that is gyroscopic and anti-

A modern Vril poster with a photo of Maria Orsic.

gravity within a closed system. The foo fighters were apparently miniature versions of these pulsing, brightly lit balls that were flying through the sky. A Haunebu contained one central plasma gyro and three smaller ones around it as directional gyros.

With the Third Reich collapsing—the failure to take the oilfields at Baku a major turning point—many Germans faced the destruction of their country and more loss of life. Some Vril Society members must have been in the parties that fled via Haunebu or Vril—maybe even in the very large Andromeda Craft. They could have gone north, to still-functioning bases in Norway, like Banak, and on to Greenland or the Blue Island in the Canadian Arctic. But in these places they would be stuck in military bases, not the ideal spot for people trying to restart their lives. They might have gone by a southern route, one that ultimately took them to South America. Here they could live a fairly normal life, with restaurants, homes, parks, movie theaters, and other recreation. And, they would be working to reconstitute the Reich in Latin America, particularly in Argentina, Chile and Paraguay.

These agents and others were now part of the Black Sun underground world that included large sums administered by Martin Bormann. There were U-boats, flying saucers, secret bases in Antarctica and elsewhere that still functioned in this extra-territorial Reich. While some of the members tried to live a semi-normal life in Argentina and elsewhere, they were always looking over their shoulders, walking down back alleys, hiding in doorways, wondering if they were being followed. They had mistresses and friends, they carried guns, they met with other Black Sun operatives in cafes, bars, city parks, or hotel rooms. Perhaps they were picked up from time to time to take part in some operation or to visit some of the bases still operational to keep them loyal subjects of the Black Sun network.

The extra-territorial pursuits of the Third Reich still lingered after the war. The creation of many of these secret bases had begun in the 1930s, as had the early experimental versions of the Haunebu. We can identify the extra-territorial pursuits of the Third Reich in these five areas:

1) The Arctic
2) Remote Islands
3) The Antarctic
4) Under Water
5) In Space

We should include parts of Tierra del Fuego, the Chilean fjords and Patagonia in there as well, as it seems that there were probably several SS bases in these areas, and U-boat bases have been suspected in the remote south of Chile and somewhere in the Andes. It seems that a large section of Paraguay is also part of

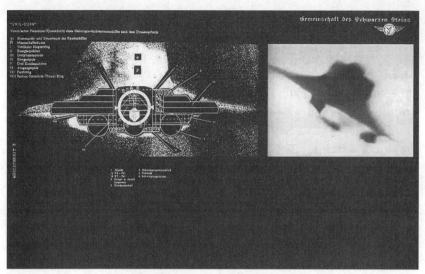

The plans for the Vril Odin craft.

Nazi-SS land holdings.

The systematic take over of Latin American countries is a theme in many of the books and articles on the SS and Nazi activities after the war. They could use the Haunebu or the Vril craft for various smuggling operations, and as vehicles for their constant psyops against the West. They had a need to keep most of these bases operational for some years after the war, but it seems that many were ultimately abandoned, especially the submarine bases. We see this in the story of the Canary Islands U-boat base.

Henry Stevens says in *Dark Star*[46] that the submarine base was fairly well known all over the Canary Islands and that many Germans had settled in the Canary Islands after the war; their families still live there today. Some Germans living in the Canary Islands during the war worked at the Villa Winter or in some other capacity, on a different island, for the Nazi cause. They stayed after the war and fellow Nazis joined them in the Canary Islands, a Spanish refuge off of the West African coast.

Similar to the Canary Islands is the nearby Spanish Sahara, a territory that would have also afforded some comfort in the form of villas and large estates with private airstrips and such. This is much as we have seen in Argentina, Paraguay and Chile where Nazi ranches have airstrips and in some cases industrial-type docks on the ocean where a U-boat could discharge people

279

and take on cargo like fresh food, clothing, and fuel. The Spanish Sahara (now part of Morocco) was only a short flight or boat trip from the Canary Islands, probably the preferred location of choice in this part of the world. Flights between the Spanish Sahara and the Canary Islands in relatively small aircraft were probably quite common during the war and in the years afterward.

The Canary Islands have also been speculated to be a site with an underwater UFO base. Numerous photos of disk-type craft have been published and appear on the Internet from the Canary Islands. The book *UFO Contact From Undersea*[3] contains an appendix on underwater UFOs (disks) coming out of the water off the coast. The book has a number of photos of the objects and one has to wonder if there is an alien base near the Canary Islands, or if it is a base for Haunebu and Vril craft plus the flying cigar-type craft that was apparently manufactured after the war in the Arctic.

The Secret Flights of Squadron 200

Aiding in the preparation for these "secret cities of the Black Sun" was Squadron 200 which made secret flights from Germany and Norway to Japan and from Europe to the Canary Islands. Stevens says that the Ju-390 flew to Japan, over the polar route, from a secret base in Norway. The Ju-390 left Norway from a base in the far north at Bardufoss, flew over the north pole, over the Bering Strait, then down the east side of the Kamchatka Peninsula to the island of Pamushiro which was within the Japanese Empire. From there it flew over Japanese-controlled Manchoutikuo to Tokyo. Stevens says confirmation of this flight comes from a radio report by the Japanese Attachés for Marine Aircraft in Germany, dated March 21, 1945. Says Stevens:

> The purpose was to transport German high-tech weaponry secrets and some personnel involved with this work. This was exactly the same purpose as the famous U-234 voyage and the two methods of communication can be considered complimentary, each part of a larger whole. There may have been and probably were more flights involving the Ju-390 to Japan just as there were probably other U-boats delivering technology to Japan besides U-234.[46]

280

A photo of a Haunebu over the streets of Tianjin, China in 1942.

Stevens says that the Focke-Wulf Fw-200 Condor was also used in the many long-range flights. This was originally a four-engine passenger aircraft but the Luftwaffe adapted it for long-range reconnaissance and even as an anti-shipping bomber. Besides extensive work in the North Sea and polar regions surrounding Norway, this aircraft also transported supplies all the way to Stalingrad in 1942.

Stevens also says that in 1944 there was a conference at Strasbourg of SS officers and other Nazi officers at which it was agreed that blueprints, machine tools, secret weapons, specialty steel, gold, money and scientists were to be transported outside the Reich for future use. The means of this transfer was the deployment of Geschwader 200 (Squadron 200), an elite flying group of young men, many of whom had lost their families. British intelligence was known to use orphans as part of their most dangerous operations (Agent 007 is said to be an orphan by author Ian Fleming). On both sides, the most dangerous missions were best done by those without families.

Geschwader 200 was the first choice of the Luftwaffe for truly

281

A photo of a Focke Wulf Fw-200 Condor with Nazi markings.

dangerous missions. Geschwader 200 was, at one point, scheduled to fly the manned version of the V-1 rocket to high value targets. This would have certainly meant the loss of almost all of the pilots, as the men themselves knew. They were, briefly, the Kamikaze squadron of the Third Reich. This project was abandoned but Geschwader 200 was involved in a suicide attack on a bridge spanning the Oder River in the final stages of the war in an attempt to slow the Soviet advance. Says Stevens about the transfer of people and goods out of the Third Reich:

> In the transfer specified in the Strasbourg agreement, the goods and people necessary were loaded aboard Condor aircraft by Geschwader 200 and flown from points within the Reich to Madrid, then on to Cadiz. Cadiz is on the Mediterranean Sea. From there transport to South America was accomplished using both Fw-200 Condors of Geschwader 200 and U-boats. Additionally, the Azores or the Canary Islands were used as a stopover base.[46]

These Focke-Wulf Condors could make long journeys, but could not fly as far as the Amerikabomber-type planes, such as the Ju-390, being developed at the end of the war. But flights to Spain and then to the Canary Islands or the Spanish Sahara were easily within their range, and the Condors could even fly from the Canary Islands to South America where they would probably land at a private airfield in Argentine Patagonia where Germans had large tracts of land.

Stevens says that the researcher/writer Friedrich Georg claimed the following tasks for the Ju-390 aside from being a long-range bomber:

1. Secret intelligence operations.
2. An escape vehicle for the Nazi leadership.
3. A long-distance delivery vehicle connecting to Japan.

Stevens mentions that the idea of giving the design for the long-range plane to the Japanese is included in number three above. The Nazis were happy to give the Japanese their technology and the Japanese would have found the Ju-390 aircraft useful, as it would have traversed the Pacific just as it would the Atlantic. Stevens also mentions that the SS commander Dr. Hans Kammler was said to have escaped Germany at the end of the war in a Ju-390, supposedly flying to South America. Alternatively, Kammler is said to have committed suicide, among various other accounts of his demise, none of which have ever been proven. Kammler is literally said to have died in a dozen different ways. Did he in fact escape to South America? Kammler was in charge of special projects for the SS and would have been quite familiar with the Ju-390. Himmler is also said to have escaped from Germany in a captured British plane flown out of a special SS airfield near Berlin. He allegedly flew to Spain and then made the journey to South America, probably by U-boat. In this version of his fate, a double of Himmler killed himself with a cyanide pill when he was captured by British troops at the end of the war, as the official story goes.

With the creation of these secret bases and repositories the Reich could regroup and perhaps make a big comeback. The British, aware of these SS psyops perhaps more than other nation, were busy doing some psyops of their own. They created a movement of rock and roll and LSD, taking the 1960s by storm and reversing a terrible trend of wars and assassinations around the world, including the assassination of President Kennedy. They turned the flying saucer invasion into Lucy in the Sky with Diamonds and the dark U-boats became Yellow Submarines.

John Arthur Reid Pepper was the top officer for British Intelligence in North America during the 1950s and 1960s and the

Beatles gave him a nod with their *Sgt. Pepper's* album, along with fellow British Intelligence agent Alistair Crowley and a number of other such agents. The person whose mind was "blown out in a car" was not Paul, but a member of British Intelligence that all of the Beatles knew. Rock and roll was banned in the Soviet Union during the 1960s and 1970s, but it was popular all over the world including in Germany, Argentina and all over Latin America where many Beatles and Jimi Hendrix-themed bars, cafes and music shops abide. Bob Marley arrived in the 1970s and an entire new culture of reggae and rock and roll took over countries in Africa, Asia and Latin America. Rock concerts are where everybody gets together to root for the same thing and a splendid time is guaranteed for all. The Beatles toured in Argentina.

Nazi and fascist ideals were defeated, as was the fear of alien invasion. Submarines were now yellow and a lot of fun. It is well known that British Intelligence and the CIA have promoted the use of LSD and other psychedelics among various populations. The main reason for this is to get people who are indoctrinated in various forms of "thought control" to envision something else from what they have been taught, often by the Catholic Church or other religions, or from political thought control such as communist or fascist indoctrination. We are all victims of thought control, as long as we interact with society at large.

This thought control extends as well into scientific realms and this includes the subject of UFOs, Haunebu craft and secret technology. Thought control and overt efforts sought to erase the name and inventions of Nikola Tesla for decades, but now his work is recognized and celebrated and the Tesla car corporation is one of the most successful and well known companies in the world today. But with the giant oil war of WWII, and the failure of Germany to take Moscow and the oilfields at Baku, where would these defeated SS fanatics go with their Haunebus—the dark side of the Moon?

Chapter 8

Dark Side of the Moon

And if the cloud bursts, thunder in your ear
You shout and no one seems to hear
And if the band you're in starts playing different tunes
I'll see you on the dark side of the moon.
—Pink Floyd, *Brain Damage*

As mentioned earlier, even before WWII began the Germans began creating special secret bases for their use outside of Germany. These secret bases were said to be in Antarctica, Tibet, Greenland, the Canary Islands, and other places. Immediately after the war there was a lot of activity in Scandinavia, mainly in Sweden and Finland, of unidentified flying objects looking like rockets or large airplanes.

These sightings were typically called Ghost Rockets or Ghost Flyers because no one knew what the objects were, where they were going, or who was flying or launching them. While some UFO authors have speculated that ghost rockets and foo fighters were related to extraterrestrial activity, others have written that this activity is related to the Nazis before, during and after the war. Like the stories of secret German bases in Greenland and the Canadian Arctic, the stories of ghost rockets and ghost flyers may be related to secret bases in northern Finland. We know that the airbase at Banak in the very north of Norway was used by long-range planes during the war. Was there a similar base in northern Finland? Remember, during WWII certain nations fought with the Axis during the war that are often forgotten. Among these nations were Finland and Romania. Both fought against the Russians on the side of the Nazis and paid a high price for this.

In his privately published book, *The Secret to Rational Space Travel*, German author Klaus-Peter Rothskugel says that secret bases were set up as early as the 1930s in northern Finland. Says Rothskugel:

> At the beginning of the 1930s, especially in the northern part of Europe, huge multi-engine aircraft with a dull painting and no national markings emerged mainly at night or in severe weather conditions and circled over cities but also over remote areas, sometimes switching on one or several search-lights.
>
> So during the winters 1933-34 and 1936-37 many such reports came from Sweden, Norway and Finland. Even the Swedish Air Force tried to hunt down these mysterious airplanes.
>
> The hunt resulted in the crash of a total of six Swedish bi-plane fighters, while the aircraft they were hunting seemed to have no problems in flying at night and in severe snow storms or in extreme coldness.
>
> There were many theories what kind of airplane cruised over the Scandinavian countries. One explanation was that the planes could have come from Germany. A certain quantity of German personnel from some German aircraft companies (Junkers) were indeed working in Sweden to test new aircraft types or to sell German technology to the Swedish military.

Ghost Flyers and Ghost Rockets

Rothskugel says that one witness who saw such a ghost flyer told a Swedish researcher:

> I am still steadfastly convinced that the airplane my father and I saw at a moonlit winter's night in 1934 was a large, low-winged, four-engined monoplane. Such did not exist in Germany at that time... I might add that my father got the outline of the airplane verified by other people who had also observed it.

286

Rothskugel says that Swedish and Norwegian military commanders had the opinion that the planes came at least temporarily from a mobile base in the west of Norway and that the numerous reports from the Gulf of Bothnia, between Sweden and Finland, pointed

A newspaper article on ghost rockets.

to bases in Finland. He says that this would indicate that these countries could have secretly agreed to use certain airfields or remote landing strips for clandestine tests.

On 30 April 1934, Swedish Major General Reutersward released the following press declaration:

> Without doubt there is an illegal air traffic going on within our restricted airspace. There are many reports of reliable persons who watched the mysterious craft very well. And there were always the same observations: All the airplanes had no national markings nor any identification numbers…

So, it seems that Finland and Norway may have had secret SS bases as early as 1934. Also, it is interesting that one of the ghost flyers was said to be a four-engine aircraft. Such aircraft are typically large bombers, but supposedly no four-engine airplanes had been built at this time—but why not?

The Messerschmitt Me 264 was the first of the long-range strategic bombers developed for the German Luftwaffe.

A photo of a ghost rocket, July 1946.

287

A Messerschmitt Me 264 four-engine bomber in flight.

The design was later selected as Messerschmitt's competitor in the Reichsluftfahrtministerium's (the German Air Ministry) Amerikabomber program, a program to develop a strategic bomber capable of attacking New York City from bases in France or the Azores.

The origin of the Me 264 design came from Messerschmitt's long-range reconnaissance aircraft project, the P.1061, of the late 1930s. These designs were also successfully used in the long-range Messerschmitt Me 261, itself originating as the Messerschmitt P.1064 design of 1937. The first flight of a four-engine Me 261 was officially on December 23, 1942 and only three were supposedly built.

But these aircraft all have their origins in long-range reconnaissance aircraft projects that do go back to the 1930s. Because of the secret nature of the war buildup in Germany in the 1930s, and all the preparations made for war on France and the Soviet Union, it seems that many secret projects were already underway in the early 1930s and that four-engine long-range craft had been built already as experimental craft. Perhaps all of these experimental craft were controlled by the SS starting in 1934. This also lends credibility to some of the Vril Society claims that the Vril craft were developed and flown in the 1930s.

Rothskugel says that in 1936 the unknown aircraft returned again on the same routes they used in 1934: Coming from far north in a southerly direction, flying over the northern part of Norway, across Sweden and back again. He says that another theory was that a Soviet heavy bomber wing, based in the Murmansk area, was conducting training missions with their four-engined TB-3 bombers or testing new equipment. Says Rothskugel:

This could indeed explain the violation of the airspace of the three Scandinavian countries but not the inactivity of these countries to solve the mystery that occurred twice (1933-34 and 1936-37, always in wintertime!).

One might assume that some special evaluation flights were conducted under extreme weather conditions, especially in cold and dark nights. These could have been secret tests of some new flight navigation systems working under these bad circumstances. And strangely the same (?) air bases might have used some ten years later, possibly again with the agreement of the three Scandinavian countries. In 1946 and 1947 this time unconventional and still unknown airplanes circling in the northern airspace, later widening their flight path to other parts in Europe.

One probable explanation could be that a certain equipment (withstanding cold and freezing conditions) with special converted multi-engined airplanes was tested in the 1930s. These operations were totally shrouded in secrecy and could have been undertaken with the participation of several nations to prepare an even more secret mission in an area where cold and icy conditions prevail and where visibility and orientation is difficult even under normal daylight conditions.

Such a field of operation could be for example the Artic or the Antarctic region. After WWII in this very Antarctic region UFOs, flying cigars and flying spheres were frequently seen and the famous Antarctic pioneer Admiral Byrd ran into trouble while on a secret mission at the South Pole (at least the official version is explaining this. Byrd, who was in Hamburg, Germany before WWII to lecture about the South Pole was probably part of a highly secret mission which is still today covered-up). Even during WWII there were rumors that Germany transferred

A pthoto of an Andromeda Craft in flight from the Ralf Ettl documents.

in secrecy personnel and equipment to the South Pole.

The flight exercises under extreme conditions conducted in the 1930s could have been preparations for a special mission, which started right after the war. A multi-national effort to begin a huge and fantastic experiment with unprecedented dimensions.

More reports of ghost rockets and ghost flyers were made after the war. On February 26, 1946, Finnish observers saw a rocket-like object in the sky that they could not explain. Was it one of the Andromeda craft that had been built? About 2,000 sightings were logged in the skies above Sweden and Finland between May and December 1946. There was a peak on 9 and 11 August of

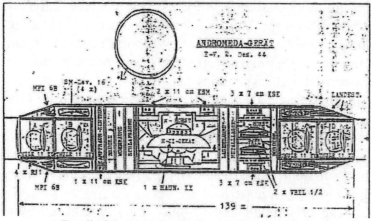

Detail of the plans for the large Andromeda Craft.

1946. Two hundred sightings were verified with radar returns, and authorities recovered physical fragments that were attributed to ghost rockets. Were these rockets that would eventually crash somewhere or were they piloted craft such as the Andromeda craft on their way to and from a secret base in northern Finland? Many ghost rocket sightings are of craft moving south, not north.

A Secret Rocket Base in Libya

Rothskugel also thinks that there was a secret research and development base in Libya during the war, with some of the personnel being transferred to the secret base in the desert from the Peenmunde rocket research center in northern Germany. He even thinks that the Germans may have tested an atomic bomb in Libya a year or so before the country was liberated by the Allies in February 1943. He thinks that a secret rocket testing sight was built in Libya even before the war. Says Rothskugel:

> Mussolini's army invaded Libya before WWII. Interestingly, Peenemuende could have used territory during the Second World War controlled by the Italian army in Libya, probably a region with secret Italian R&D installations for special tests (rocket launches, A-bomb tests, etc.).
>
> Concerning North Africa, there is a German intelligence report, allegedly once written by Wernher von Braun and

his group. The "Bericht zur USA-Rakete Comet", Report on the US-rocket "Comet", dated 6 August 1943 is telling most probably an invented story about a certain German engineer or technician working for Peenemuende. He was captured by British troops somewhere in North Africa. This rocket specialist got later as prisoner of war to a POW camp in Nazareth in Palestine.

A German rocket specialist in North Africa, what had he to do over there? From which place, installation etc. did he come from? To what special group did he belong? What was his mission in Africa?

The very report dated August 1943 was signed by a certain Gefreiter Guenter H.H. Goetsch and contains a lot of facts and names which were only known by insiders and totally unknown in the public, especially in Germany, because these highly secret information were never published before or during the war. This is another indication that Wernher von Braun could have been responsible for writing the report completely by his own.

The faked story seems to be prophetical, because an U.S. interrogator asked the "Private First Class" Goetsch, if he would like to come to the USA after the end of hostilities, to go on with his research work, this time for the U.S. space industry. Exactly as did von Braun after WWII!

So it could be possible that from 1943 onwards Wernher von Braun already knew that he will meet at the end of the war certain people like Richard William Porter, a specialist in electronics. Dr. Porter had a leading position within the U.S. electronic company General Electric and he was later involved in "Operation Hermes."

Rothskugel may be on the right track here, and it may be that SS operatives were still in Libya after it had been taken by the Allies in February of 1943. This may explain some of the odd UFO sightings in North Africa in the years right after the war. These sightings probably involved Haunebu craft and will be discussed shortly. Asks Rothskugel about the activity in Libya:

...Could it be possible that Peenemuende was also active in North Africa, i.e. in Libya? There is a story, that some POWs of the German "Afrika Corps" had a retouched photo or post card in their bags, showing New York City with a huge mushroom developing over the city. A factious impression of a German attack with an atom bomb!

Did the Germans conduct together with Italian specialists atomic tests in occupied Libya until the Allies liberated North Africa? What could have happened to these secret installations somewhere in Libya after the Axis Powers were gone? Did the research go on, now under control of Allied special groups? Something such as this may have been done in the Jonas Valley after WWII?

...But there is another hint that Libya was involved in certain nuclear tests or in atomic explosions, perhaps together with even more fantastic experiments:

The LDG, the famous "Libyan Desert Glass" was used long ago to make knives and sharp-edged tools as well as other objects. Two areas exist in the Libyan Desert where atomic explosions could have taken place in ancient times and which are still visible today. One is oval-shaped, the other more a circular ring. Did an atomic explosion in that region, thousands of years ago melt the sand to glass?

Wrecked Italian aircraft at the destroyed airfield in Tripoli, 1943.

293

So, Rothskugel muses that the Libyan Desert Glass may be both ancient and modern. Perhaps he thinks it a coincidence that an ancient atomic detonation was near the site of a recent atomic detonation, a notion that history is cyclical. If it was recent it would show more radioactivity then it shows now, I think. Still, it is not necessary for this glass to be recent for there to have been a secret rocket research base in Libya somewhere. If this base did exist it may have been visited by Haunebu craft after the war and contributed to some of the strange UFO sightings that were occurring all around the world after WWII. We already talked about some of the early sightings starting in 1947. It turns out that the CIA was following these sightings around the world and keeping quite a file on them.

Haunebus All Over the World

In his book *The CIA UFO Papers,*[40] investigator Dan Wright says that CIA reports from 1953 indicate that there was a great deal of UFO activity around the world that year. He says a document refers to the Danish Defense Command remarking that the "flying saucer traffic" over Scandinavia was of immense aero-technical interest.

Another CIA paper dated August 18, 1953 conveyed accounts of anomalies from newspapers in Athens, Greece; Brazzaville, Congo; and Tehran, Iran. Wright then mentions another CIA paper from that year that said:

> ...a German engineer claimed that flying saucer plans, drawn up by Nazi engineers before World War II's end, had come to be in Soviet hands. The source claimed German saucer blueprints were already underway in 1941. By 1944, three experimental models were ready, one in disc shape. All could take off vertically and land in a confined space. After a three-month siege of the German's Breslau (now Wroclaw, Poland) facility at the war's conclusion, Soviets stole the plans on saucer construction.

Wright then describes another curious CIA document with an account from November 22, 1952:

A missionary and five companions in French Equatorial Africa had had a close encounter. Driving at night, they had witnessed four motionless discs overhead that lit up like suns when in motion but were silvery when stationary. Over 20 minutes the four moved about the area, seemingly performing tricks, then hovered momentarily before leaving non-uniformly. Later the six witnesses saw four objects forming a square at cloud level. One lit up vivid red and rose vertically; the other three joined it to form a square again. Luminous aerial objects were seen in the same time period above Homs, Syria, and the oil fields at Abadan in west-central Iran.[40]

Wright mentions two strange cases in Africa from 1952 in the CIA documents. One incident took place on June 1, 1952 when the master and first mate of a cargo ship just off of Port Gentil, in the West African country of Gabon, witnessed an orange luminous object rise up behind the port, do two right-angle turns, pass overhead, and continue out of sight. Was this a Haunebu picking up or letting off passengers in a field behind the port?

This craft might have then departed for Laâyoune (or El Aaiún) that at the time was the capital of the colony of Spanish Sahara. It became a modest city in the 1930s and in the year 1940, Spain designated it as the capital of Spanish Sahara. Because Spain was a neutral country, but pro-Nazi (like Argentina), its colonial lands were used as proxy territory by the Third Reich, and U-boats and aircraft made routine stops at locations where supplies were available and transactions could take place. The country was disestablished in 1976 and today is part of Morocco. As has been noted, one early report said that Hitler was living in Laâyoune immediately after the war.[33]

Wright then discusses a CIA document on UFO activity in July of 1952 in Algeria. Starting on July 11 in Lamoriciere, several UFOs were seen including a longish, fiery oval. Then four days later a flying saucer was seen at the town of Boukanefis. Then on July 25, at 2:35 in the afternoon, UFOs were seen over a factory and other places near the coastal city of Oran. The same night near the Algerian town of Lodi, southwest of Algiers, several UFOs were seen.

The next morning at 10:45 in broad daylight in Tiaret, Algeria, five persons saw a shining cigar-shaped UFO with a darkened center silently traversing the sky. A similar UFO was seen later that night in the Algerian town of Eckmuhl. The CIA documents say that a few days later, on July 30 in Algiers, a black disk was seen by a woman to descend from the sky and then suddenly move horizontally out of sight. Later that night, in the early hours of July 31, UFOs were seen in the sky by multiple people around Algiers, the main city of the country, and in other areas like Oued and Tlemcen. Finally, later that morning in broad daylight (11:30 AM) a couple driving just outside of the major city of Oran saw a UFO cross the sky at great speed.

So with these last sightings in Algeria in 1952 we seem to have a Haunebu or Vril craft along with a cigar-shaped craft. Did the secret rocket base in Libya have something to do with these sightings in Algeria? Did this craft depart for a secret base in the Spanish Sahara or nearby? The postwar activities of the SS were being carefully followed by the CIA and probably Naval Intelligence. While we have been able to see many of the CIA and FBI files concerning UFOs because of the Freedom of Information Act, this does not apply to files from Naval Intelligence which would no doubt show much more than these CIA files do. I would like to remind the reader again that virtually only the United States has a method for citizens to request former Top Secret files on certain topics.

Meanwhile, Back in Argentina

Meanwhile, back in Argentina there was plenty of Haunebu activity going on in the 1970s, including flying saucers coming out of the ocean. In *The CIA UFO Papers,* author Dan Wright mentions briefly a curious incident when a young man saw a flying saucer move silently over pastures near the Argentinian port city of Bahia Blanca on September 19, 1971. Says Wright:

> The object stirred up whirlwinds of dust, while cows appeared to change color in its presence. Afterward, his

face was severely burned and he suffered from a persistent migraine headache.[40]

In the book *UFO Contact From Undersea*,[3] Wendelle Stevens has an interesting appendix that discusses unusual sightings, all in Argentina, of disk-shaped craft emerging from under water and flying away. He mentions how he was investigating a case where an abductee from Rio Salado in northern Argentina was told by the occupants of the flying saucer that they their base was under water. Asks Stevens in the 1982 book:

Is there, in fact, a submarine UFO base in the South Atlantic off the Patagonia Coast of Argentina? More and more incidents seem to point that way.

Ever since ten years before [in] February 1960 when the Argentine and United States Navys' combined forces tried for two weeks, with uncounted tons of explosives, to bring two unidentified objects maneuvering under the waters of Golfo Nuevo, an area of only 20 by 40 miles, to the surface, reports of UFOs in that area have continued.

Near the end of June 1950, Sr. Romero Ernesto Suarez, walking the coastal road between Rio Grande and San Sebastian in the Territory of Tierra del Fuego, late, about 23:00 hours, suddenly heard the sound of turbulent water, which increased and became more violent. He was puzzled and became a little frightened. There was no wind, storm, or water currents that could explain this noise. He peered into the darkness in that direction, and suddenly a huge, luminous, oval-shaped object emerged from the sea about 500 meters from shore. It ascended vertically to a certain altitude, made a quick 90-degree turn and disappeared rapidly to the northwest, toward Rio Gallegos.

Fifteen days later, again at night, when he was between Rio Gallegos and Santa Cruz, ascending a hill near Puerto Coyle in the province of Santa Cruz, he witnesses a similar occurrence. This time four small luminous domed discs surged up out of the water vertically, in perfect formation, leveled off and flew up the coastline a ways and then turned

An illustration of a flying saucer coming out of the water, June 1950.

left in the direction of the Cordillera de Los Andes.

Two years after that, and even 2-1/2 years after the February 1960 case so well reported in *Flying Saucer Review* Vol. 6, Number 3 and Vol. 10, Number 4, UFOs were seen to enter and also to leave the water again. In August 1962, Sr. Vicente A. Bordoli, a truck driver from Mar del Plata, while in the South, driving along National Highway Route #3 bordering the Atlantic Coast, with his son Hugo, observed strange formations of lights in the sky that entered the water in the Golfo San Matias, Province of Rio Negro. A few minutes later they emerged and ascended into the sky and disappeared. San Matias is a deep gulf with 500-foot depths in some places.

Two hundred miles south is another deep gulf, the Golfo San Jorge. On 28 July 1964, at 21:24, the subprefectura at Puerto Madryn, Chubut Province, received a message from the Argentine tanker "Cazador," saying that at 21:10 the captain and crew had sighted a strange light that fell into the water. They had plotted its bearings as latitude 45 degrees, 56 minutes and 06 seconds South and 64 degrees, 00 minutes and 00 seconds West longitude. They and another tanker, the "San Antonio" searched a five-mile circle looking for survivors and debris. In the afternoon of the next day, the Norwegian ship "Sumber" arrived in port and the captain reported to the authorities, "Yesterday at about 21:10, as we were approaching the Argentine coast, we observed the fall of an aerial object or small comet (into the waters of the Gulf). It came from the northeast, horizontally, towards shore. It radiated a brilliant light."

Two months later, in the same Golfo San Jorge, a public personality known for his honesty, who prefers to remain anonymous, declared that on 20 September 1964, while driving at night to the city of Comodoro Rivadavia, as he was leaving the town of Caleta Olivia, northbound up the coastal highway, he saw four small luminous lens-shaped objects describing a parabola in the sky as they flew in perfect formation. Suddenly they dipped, and still in formation plunged into the waters of the gulf and disappeared. He stopped his car and got out but could see nothing more.

A little later, while driving north of that position, he saw four similar luminous objects, possibly the same ones, emerge from beneath the sea, veer in his direction as they accelerated, and climb out at a steep angle at prodigious speed and disappear into space above.

Also in the Golfo San Jorge, on 18 March 1966, Sr. Carlos Corosan, a well-known resident of the area, walking the beach 15 kilometers north of Puerto Deseado, at 16:00 heard a strange rumbling sound. Looking up, he saw a strange cigar-shaped craft 18 to 20 meters long, moving about 30 meters above the waves offshore. A grayish vapor was coming from the rear of the object. It was dark gray colored and its surface shined with a metallic finish. The finish was completely smooth with no wings, windows, ports, or breaks of any kind and no markings. Suddenly it stopped ten meters above the water, commenced to vibrate with the rumbling sound increasing in intensity, and then he heard a muffled explosion. The smoke from the rear became denser and very black. The nose came up and it began a slow ascent to the north northeast, though visibly erratic. He heard another explosion followed by a reduction in the rumbling, and it descended rapidly and entered the water with a splash and disappeared beneath the waves.

Farther north, off Buenos Aires, in front of the Pinamar District, at night on 31 May 1971, psychologist Zulema

Bruno, driving her car along the shoreline road, saw a strange luminous lens-shaped object surge from the sea as it rotated on its vertical axis and radiated orange rays of light. It followed her automobile for some 300 meters and then, putting on a burst of prodigious speed ascended and disappeared into the sky above.

Stevens then discusses the amazing photograph which is also printed in color on the dust jacket of the book and shown here (see the color photo in the color section). The photo shows a dark disk-shaped object emerging from the ocean while two young men stand on a beach in front of several dead seals. Says Stevens:

On 2 December 1971, at Punta Norte, on the Valdez Peninsula, in the Golfo San Matias, Province of Chubut, at 15:00, mid-afternoon, Sr. Ricardo Jorge Espindola and several companions were hunting seals in the cold clear deep waters off the north cape of Punta Norte. They had taken a number of seals and he was shooting a picture of two of his friends with some of the seal carcasses. He had color slide film in his camera. Suddenly the water about a quarter of a mile off shore began to roil violently and a large circular craft emerged from the turbulence coming directly towards them, and then curved away and flew out to sea and disappeared. He snapped the picture and lowered the camera in stunned surprise. The object was far away before he recovered his composure and remembered to try for another. He decided it was too distant and did not take the second picture. The object was circular and of a dark color, and was closely surrounded by a reddish haze or halo effect that gave the whole craft a reddish cast. The object was soundless and left no trail.

The photo is remarkable and the many stories of UFOs coming in and out of the water in Argentina, mainly in the large Patagonia region, are credible. While UFO researchers for years believed that these incidents must be of extraterrestrial craft exiting or entering the water, but we now know that the Germans were active in

Antarctica and Argentina after the war and they designed and built craft such as the Haunebu and Vril saucers that could function as aerial craft and submersible craft.

Stevens was partially aware of German flying saucers when this book was written in 1982, but it wasn't until after 1989 that he collected the many plans and photos of the Haunebu and Vril craft. In fact, Wendelle Stevens put nearly all of these photos and plans onto CD disks that he burned one by one and sold at UFO conferences. I purchased one of these CDs from him at a UFO conference in Laughlin, Nevada circa 2004. Wendelle was a very approachable person and I chatted with him about the Haunebu and Vril photos; he believed them to be authentic and very important. I think that Stevens would have agreed with me that some of the reports above are likely to be of Haunebu or Vril craft. By the way, Wendelle Stevens (deceased in 2010) and Henry Stevens, also quoted extensively in this book, are not related.

Wendelle Stevens gives us one last curious encounter in Patagonia, this one in 1974, which was also photographed (see the color section for the color photo):

Two hundred and sixty miles south of the Valdez Peninsula, along the coastline of Golfo San Jorge, which has figured prominently above, another new case developed. At 19:40 on 23 March 1974, near coastal highway Route #3 just before reaching the town of Caleta Olivia, Sr. Cesar Elorda saw a silvery metallic domed disc approaching from the east, from the Golfo San Jorge. It was moving west on a steady course at low altitude and he could see it very clearly. He had time to get his camera ready and take one beautiful color picture on Ektachrome film. It was completely symmetrical and smoothly finished in a low conical form almost like a "coolie hat." It had no projections, ports or windows of any kind and no markings. The dome rose in a smooth unbroken curve from the flange of the disc to the top, and was smoothly rounded. It flew almost directly overhead and to the west into the sunset sky. By the time he got his camera ready it had already passed over and he shot the picture at an angle

The photo of a saucer craft coming suddenly out of the water taken by seal hunters along the coast of the Peninsula de Valdez in Chubut Province, Dec. 2, 1971.

into the sunset. He could still see the silvery finish when he snapped the picture but when the slides came back he was surprised to see that it had photographed completely black.

It is interesting to note that in several of the cases the flying saucers came out of the water, which is the eastern Atlantic coast of Argentina, and then flew west towards the Andes. It seems that there is a Haunebu base somewhere in the Argentine Andes, south of the towns of Esquel and Bariloche, where Hitler allegedly lived.

Somewhere in these bleak, snow-covered mountains is a hidden base, probably inside a mountain, where Haunebu, Vril and other craft are stationed. In my book *Antarctica and the Secret Space Program*[10] I propose that there is a submarine and UFO base run by the SS in the Antarctic Peninsula (also called Palmer Peninsula, Graham Land, or Tierra de O'Higgins in Chile). This peninsula has been claimed by the United Kingdom, Chile, and Argentina.

The peninsula forms an 800-mile (1,300-km) extension of Antarctica northward toward the southern tip of South America. The peninsula is ice-covered and mountainous, the highest point being Mount Jackson at 10,446 feet (3,184 meters). Marguerite Bay indents the west coast, and Bransfield Strait separates the peninsula from the South Shetland Islands to the north. Many other islands and floating ice shelves lie off the coast. I surmise in the last chapter of my book that a secret base is actually located on the island of King George in South Shetland Islands off the northwest coast of the Antarctic Peninsula.

This island is actually one of the closest islands in Antarctica to the southern tip of South America so it is well located in regard to Argentina and Chile, and it can be a remote station while still being fairly accessible to flying craft or submarines. It is likely from this base, probably created during WWII, that the flying saucers (and maybe submarines?) seen at Trindade Island—the former Brazilian-American submarine base—had come.

On top of the mysterious Haunebu base in the Argentine Andes of central or southern Patagonia, we have the curious case of the Cerro Uritorco near the small town of Capilla del Monte in northern Argentina, where we have already mentioned many flying saucer sightings have been made.

Ride in a Haunebu Anyone?

During the 1980s in Argentina the UFO activity centered around the previously mentioned northern city of Capilla del

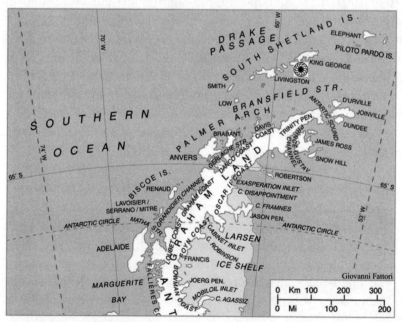

A map showing the Antarctic Peninula and the South Shetland Islands.

Monte. As mentioned earlier, Capilla del Monte is a small city in the northeastern part of the province of Cordoba, where the main tourist attraction is the Cerro Uritorco, a small mountain only three kilometers from the city, famed around Argentina as a center of paranormal phenomena and UFO sightings. One famous incident happened on January 9, 1986. On that day a flying saucer was seen near Capilla del Monte by hundreds of residents. The astonished crowd watched the craft land on a nearby hill known as El Pajarillo. Later, residents rushed to the scene where they found a mysterious footprint on the ground where the flying saucer had landed. As we reported in an earlier chapter, the area around Capilla del Monte has been a major UFO area since the 1950s. There has also been a lot of activity in the area by "Men in Black." It seems likely that both of these mysterious occurrences are related to postwar German activity in the area.

About three years later on the sunny afternoon of December 26, 1988, a silver UFO flew over the northern Buenos Aires suburb (barrio) of Villa Urquiza. It is considered the most spectacular UFO incident to occur in Argentina, with more than 7,500 witnesses. The local airport reported an object flying to the west, towards General Paz Avenue, seen on the radar.

In 1991 a major flap of UFO sightings occurred around the town of Victoria, a city located in the southwestern part of the province of Entre Ríos, Argentina, just west of Buenos Aires. Victoria is located on the eastern shore of the Paraná River, opposite the major city of Rosario, which is in Santa Fe province. The Paraná River here is quite wide and commercial fishing is a major activity in the area.

Witnesses Andrea Pérez Simondini and his wife Silvia wrote in 1998:

> Upon reaching the area, we installed ourselves and our gear at the Victoria campgrounds, since it was one of the best observation spots facing the lagoon, dominating every detail in the island sector, from which the maneuvers of these objects could be witnessed constantly. At the edge of one of the canals bordering the campgrounds, we saw a mustard-yellow light toward the north, rising over the

horizon and remaining stationary for some 8 minutes, as if hanging from the sky. The possibility that it was an antenna we hadn't noticed earlier was discussed, or had it really been an object? Momentarily, faced with disbelief, I stared at the light, which after a few minutes began traveling toward the northeast, that is to say, toward the center of the city.

I began signaling it with a flashlight, and it suddenly changed its trajectory toward me. I quickly went to where the others were watching, as they had also noticed the change in movement.

When the object crossed a row of trees that runs along the edge of the camping, we noticed that it had lights facing forward. As it flew overhead, those lights cast their beams over us as though looking downward. In a matter of seconds, the object turned off and vanished as if it had never been there. Our astonishment lasted for hours—we could not believe such a thing could have happened to us. I think the story conveys the sensation that engulfed all those who were witnesses to this incredible experience.

...We interviewed direct witnesses to the objects' maneuvers, those who had found possible evidence of them, those who claimed having seen entities or beings emerging from craft that landed on the fields. A wealth of information allowed us to establish some significant statistics. But the substantial element was that these objects—contrary to what was believed—did not emerge from Laguna del Pescado. Rather, they emerged from one of the branches of the Victoria River, a brook with a strong current. This brook was the deepest area, and eyewitness accounts not only described the lights that emerged or came out of there, but also the noises that could be heard under the water. We stationed ourselves there several times and on several occasions were auditory witnesses to the sounds that caused great fear among the locals.

Regarding the objects, they said that their luminosity was so great that many of their homes had dark drapes over the windows, since the objects could bathe the ranches in light. It is somewhat humorous to see all these little houses with black curtains.

So, what of this amazing flying saucer light show over an area not too far from the capital megacity of Buenos Aires? It appears that the craft, seen coming out of the water at times, made deliberate passes over ranches and other areas with its powerful lights on full blast to illuminate people's houses in the middle of the night. Why would sneaky extraterrestrials be so overt about their activities in this area? They wanted people to know that they were there. Were they trying to frighten people? It doesn't seem like it. Were they completely unafraid of the Argentine authorities? It does seem that way. Were they trying to be noticed in order for people to realize that the flying saucer phenomenon was real, and not just a fantasy coming out of the media?

Indeed, having spent years travelling around South America, starting in the mid-1980s, I have found that people on this continent have a high degree of belief in a real UFO phenomenon. People on the street, the media, and even the governments, all largely admit that UFOs, called OVNIs in Spanish, are very real.

There is just too much UFO activity in South America to cover exhaustively here. By this time the reader should have a pretty good idea of just how strange things have gotten in Argentina, Chile, Brazil and other countries since the end of the WWII. Much of this activity seems to be from a remnant of the SS and the Black Sun group with their Haunebu and Vril craft.

Hans Kammler and the Black Sun Survival

Hans Kammler was a high-ranking Nazi and an SS-Obergruppenführer, the highest rank in the SS, and one of the most powerful men in the Third Reich right up until the end of the war, and probably beyond. Kammler joined the Nazi Party (NSDAP) in 1931 and held a variety of administrative positions after the Nazi government came to power in 1933, initially as head of the Aviation Ministry's building department. He joined the SS in May of 1933. In 1934, he was a councilor for the Reich's Interior Ministry.

Kammler was also charged with constructing facilities for

various secret weapons projects, including manufacturing plants and test stands for the Messerschmitt Me 262 and V-2. Following the Allied bombing raids on Peenemunde in Operation Hydra in August 1943, Kammler assumed responsibility for the construction of mass-production facilities for the V-2. He started moving these production facilities underground, which resulted in the Mittelwerk facility and its attendant concentration camp complex, Mittelbau-Dora, which housed slave labor for constructing the factory and working on the production lines.

In 1944, Himmler convinced Adolf Hitler to put the V-2 project directly under SS control, and in August Kammler replaced Walter Dornberger as its director. From January 31, 1945, Kammler was head of all missile projects. During this time he was also partially answerable for the operational use of the V-2 against the Allies, until the moment the war front reached Germany's borders. As an SS officer, Kammler was the last person in Nazi Germany to be appointed to the rank of SS-Obergruppenführer. In March 1945, partially on the advice of Goebbels, Hitler gradually stripped Goering of several powers on aircraft support as well as maintenance and supply while transferring them to Kammler.

Kammler is listed as dead on May 9, 1945, the day that he disappeared and one day after the official surrender of the Third Reich. Kammler essentially vanished into thin air with numerous stories told in later years to explain his death or disappearance. He was variously said to have: committed suicide by taking a cyanide pill and then to have been secretly buried; been captured by US troops and brought to the US; continued to live in Austria or somewhere else in Europe; fled to South America. One might

Hans Kammler, center, in an SS uniform.

imagine that Kammler was on one of the flights to Tibet during or after the war.

Henry Stevens says that the SS-Obergruppenführer Hans Kammler had probably been part of a planned coup to overthrow Hitler in what was known as Operation Avalon or Schwarze Adel, so he might have been considered to have been part of the "resistance" if one were willing to stretch the term.[46]

He also says that an eyewitness, a German engineer named Hans Rittermann, spoke with Kammler in his new position long after the war ended, and Kammler was not in hiding in South America but in Prague, sustained by both the East and West and working for the Americans and the Soviets. He said that Kammler had cut a deal with both sides.

Hans Kammler in an SS uniform.

Like Wernher von Braun, Kammler must have sent intermediaries ahead to cut a deal before his actual surrender, and he must have had a plan to disclose some facet of the technology he was charged with developing as a bargaining tool. This might have been, for instance, sending someone ahead to divide the file in question into two parts says Stevens.

Kammler had worked and made his career in what is now the Czech Republic. Kammler spoke Czech. Prague is a world of its own, neither fully Slavic nor German, but an island in between. This was a place where both the German and Czech languages could be heard on the streets. Kammler was home! He must have been given comfortable surroundings, and a good allowance or salary.

Stevens says that Kammler, now back in Prague dealing with both the Americans and Soviets, must have had connections to Otto Skorzeny and the Rheinhard Gehlen Organization with its program of bringing important Nazi scientists to the US in Operation Paperclip. Stevens says that Kammler would have been able to get word out to his former associates in the SS and he quickly became part of the Nazi International, an extra-national group with interests all around the world.

Stevens says that Kammler supplied high technology to both the East and West in exchange for the technology that they had developed on many fronts. Says Stevens:

> The Nazi International would have been able to move around the planet in their flying discs and mystery U-boats at will. The Gehlen Org. would have given them a "heads up" to any problems and supplied them with insider, secret information about both the East and West. Gehlen, by some accounts, stoked the fires of the Cold War, inflating some intelligence to keep the confrontation going. This all worked together so well. In a way, Kammler had his Fourth Reich and he was Fuehrer.
>
> The axis of this "Fourth Reich" or Third Power as it should be known, was Madrid (Skorzeny), Munich (Gehlen), and probably the Prague/Pilsen area for Kammler that he knew so well.

The workshops were in South America, being supplied with sub-assemblies and parts for "Mimes Schmiede" most from the USA. Money was no problem since Dr. Hjalmar Schacht was alive and well and the Third Power was probably making money on its own on an ongoing basis. This trade in Nazi science and technology and the relationship with both the East and West probably had its ups and downs but went on until the early-mid-1970s when Kammler died (1972) and Skorzeny died (1975). This was about thirty years after World War Two and the older generation must have been considering retirement. With the coming of the 1980's, there must have been some sort of change of leadership and perhaps new goals and outlooks.

So apparently Kammler lived for decades after the war, and probably even took a flying saucer ride or two, but always had to live in hiding. Like Rheinhard Gehlen, Otto Skorzeny was able to live his life in partial exile in Madrid and did not have to go into hiding. In the years after the war, with the defeat and division of Germany, loyal Nazis must have had a difficult time trying to defend a defeated ideology. If they were to suddenly get a ride in a Haunebu, perhaps in South America, wouldn't that be a thrill! Your faith that the Reich would be restored might get a small boost from such a demonstration of the genius of the Third Reich, despite the fact that they had been defeated and humiliated by Russia and the West.

Skorzeny was a tall, Austrian-born SS-Obersturmbannführer (lieutenant colonel) during World War II. During the war, he was involved in many operations, including the removal from power of Hungarian Regent Miklós Horthy and the famous rescue mission that freed Benito Mussolini from captivity in a mountaintop castle.

Skorzeny was held in Germany by the Allies for two years but he escaped from an internment camp in July of 1948 with the help of three SS officers disguised as American guards. He hid out on a Bavarian farm for 18 months, then spent time in Paris and Salzburg before eventually settling in Francoist Spain where he was given asylum. Though Skorzeny was unable to travel to certain countries, like Germany, Austria, Britain or the United States, he was essentially a free man and could travel freely with a Spanish passport to many countries, including Ireland, Switzerland and all of South America.

Starting in 1950 Skorzeny traveled frequently between Spain and Argentina, where he acted as an advisor to President Juan Perón and as a bodyguard for Eva Perón, while fostering an ambition for the "Fourth Reich" to be centered in Latin America.

In 1953 he became a military advisor to Egyptian President Mohammed Naguib and recruited a staff of former SS and Wehrmacht officers to train the Egyptian Army, staying on to advise President Gamal Abdel Nasser.

In 1962, Skorzeny was allegedly recruited by the Mossad and conducted operations for the agency against missile scientists in Egypt.

In the 1960s, Skorzeny set up an early group of for-hire mercenaries called the Paladin Group, which he envisioned as "an international directorship of strategic assault personnel [that would] straddle the watershed between paramilitary operations carried out by troops in uniform and the political warfare which is conducted by civilian agents." They were based near Alicante, Spain, and specialized in arming and training guerrillas. Some of its operatives were recruited by the Spanish Interior Ministry to wage a clandestine war against the Basque terrorist group ETA.

Skorzeny had homes in Ireland and Majorca, as well as in Madrid. He may have had property in Argentina, though he probably didn't really need it. His friends had plenty of land and even flying saucer factories. Officially, Skorzeny died of lung cancer on July 5, 1975 in Madrid at the age of 67. His funeral in Madrid was attended by a number of former Nazis as well as Mossad officers. Some writers have suggested he lived on past this funeral, and finished his final days at an estate in Florida where his daughter lives to this day.

We can see that the trio of Rheinhard Gehlen in Austria, Skorzeny in Madrid and Kammler in Prague would be a strong axis of evil to control a worldwide empire of clandestine businesses and factories. But we never learn who the chief officer in South America was. Perhaps it was Skorzeny who was constantly moving

Reinhard Gehlen in 1975.

between Spain and Argentina in the early years after the war. Or, perhaps our mystery man—the Dr. Evil of South America—was someone else, someone who remains anonymous today, though it is likely that such a person is dead by now, having died in the 1970s or 1980s. It has been said that Juan Peron sold Martin Bormann ten thousand Argentine passports in 1947, so perhaps it was Bormann who was "our man in Argentina." Bormann is officially listed as killing himself with a suicide pill in 1945. However, it seems he escaped to South America and controlled the massive amount of Nazi money now in Argentina.

But in any case, Skorzeny was apparently in constant contact with Kammler and Gehlen and would then fly to Argentina—on a commercial flight—and meet with his many Nazi contacts in that country. From Argentina Skorzeny may have visited the base in Antarctica, probably in a Haunebu.

The Final Integration of the Third Power

The evidence seems clear that WWII did not come to an end with the surrender of Germany and Japan. It has been pointed out earlier that the SS never actually surrendered at the end of the war, and some claim that Japan never officially surrendered either. Unlike Germany, which was overrun by Russian, British and American forces, Japan was never overrun by the Allied forces, but rather two of its major cities were "nuked" and thus Japan capitulated to the Allies. At the time of capitulation Japan essentially still occupied all of the territory that it had taken during the war. It took nearly a month for Japan to pull its forces back to Japan, leaving a massive power vacuum in many areas formerly under its control. Many of these countries, such as North and South Korea, are still dealing with this sudden end to the fighting in the Asian theater; these countries began to have their own fights inside their own countries—divisions promoted by the Americans on one side and the Russians on the other.

The Black Fleet with its secret SS labs and secret cities in South America—and even Antarctica—managed to last for decades as a Third Power that controlled whole countries in South America at times. We can easily put Argentina, Chile and Paraguay into this category, and other countries such as Brazil, Bolivia and Peru were heavily influenced by this Third Power that apparently used secret saucer bases in the mountainous areas.

Stevens says that the Third Power ultimately came to a truce and made an agreement of sorts with the United States military during the second term of President Ronald Reagan. Stevens says that about the same time as his famous "Mr. Gorbachev, tear down this wall" speech on June 12, 1987, Reagan made a side trip to an obscure town in far western Germany, a town called Bitburg. At a small ceremony Reagan placed flowers at the graves of the German soldiers who had died there defending their country from the invading Western Allies near the end of the war. There are no Americans buried at Bitburg, but there are a large number of Waffen SS (the military branch of the Nazi Party's SS organization) soldiers buried at the small cemetery. These SS officers had fought against the British and Americans at the end of war.

Why would President Ronald Reagan do this? Stevens suggests that this was a formal integration of the former Waffen SS and its many tentacles into the US military's secret space program. President Reagan also ordered that all Pershing missiles based in Germany be removed. Reagan had a very pro-German foreign policy while at the same time offering overtures to the Russians, who also wanted the Pershing missiles removed from Germany.

The United States was at this time given some special technology by the Third Power, Stevens claims, including maser and laser technologies that the United States didn't have. They also gave the Americans klystron technology that can be used to produce positrons for space travel. We might think that some of the space technologies were developed at the secret SS labs in the Arctic, Antarctica and South America—and possibly even Tibet. Earlier we discussed briefly that the SS still maintained laboratories in Western Germany well into the 1950s. No such secret SS labs would have been possible in Eastern Germany, it would seem.

Says Stevens about klystron anti-gravity technology in which beams of positrons are projected above or below a discoid craft:

> Then let us suppose a flying craft powered by positrons generated via klystron technology. How would this look? Well, how about mounting the klystron-positron projectors as swivel heads, mounted on the wing tips of a flying triangle? At the point where the three beams converge, that is the point into which the flying triangle moves, pushed and pulled by this super Biefield-Brown Effect. At low power, for instance, with the three beams converging over the center of the craft's gravity, it would just hang there, motionless. With variations in power and focus, any type of three-dimensional movement is possible.
>
> If this were the technology the Reagan Administration had received, it would account, temporally, for the appearance of flying triangles in Belgium only a couple years after "payment" had taken place. Likewise, beyond Belgium this new shape seemed to be the predominant UFO shape from that time to the present.[46]

This was also a time period when the final officers of the Third Reich, including Operation Paperclip Nazis like Wernher von Braun were dying of old age. Says Henry Stevens:

> And this time-line ties into what was happening internally within the Third Power itself. The Third Power remained intact until sometime around the 1990's, in the technological sense. Wilhelm Landig had kept in contact with expatriate Reich insiders who informed him of the situation. The reader will remember the "Ruestungsesoteriker," the armament esoterics,

the ultimate geeks who kept the postwar Nazi high technology, including the field propulsion flying discs in good operation. They worked out of these secret bases and through a network supplying parts and know-how toward the perpetuation of these technological wonders which resided in huge underground caverns connected to the outside world by tunnel-entrances. Well, it seems even they had limits. All machines eventually wear out.[46]

Stevens then tells us that the German SS officer and novelist Wilhelm Landig commented on the postwar situation through an informant, Heiner Gehring, who managed a peek into Landig's private files:

Actually, after the war, there were efforts made to build a military-technical power outside of Germany. The Base 211 in Neuschwabenland, which really was established, was certainly given up after some time. Besides the immunity situation it may have also been the dropping of the atom bomb of the USA during the Geophysical Year 1954 as a reason for the handing over of the base. Garrison and materiel of the Base 211 were sent to South America. Likewise, more German U-boat bases were available.

German flying discs, so the Landig particulars let it be known, are still warehoused in South America but perhaps may not be flight worthy any more. Their propulsion was unconventional. The development of the "Haunebus" through the "Vril" Society did, in collaboration with the SS, actually take place. Certainly some of those topics that are found in the video-films in circulation are partially completely humbug. False are the reports concerning moon and Mars flights of German flying discs, the equipment was not developed for such effort. In the USA and USSR, attempts after the Second World War were made to replicate flying discs. This replica has, on propulsion, failed, which can not be reconstructed.[46]

So Landig admits that the Antarctic base, as well as the bases in the Arctic, were probably abandoned after they were "nuked" in 1954. However, it would seem that other bases in Antarctica continued to operate well into the 1960s and the secret cities in South America may well still be active to this day. Likewise, the secret airbase in Tibet is likely to have been abandoned by the early 1960s as the Communist Chinese extended their power into even the remotest corners of Tibet. Landig mentions that the craft are warehoused in South America, and may be unusable. This does not seem to be the case, as South America continues to be a hotbed of UFO stories and sightings and it would seem that craft of different sorts, even newer models, are still being flown in South America as well as Antarctica.

Landig is also probably incorrect when he says that Americans were unable to reconstruct the field effect propulsion that propelled the Haunebu and other craft. It seems likely that the Americans, who had allegedly teleported a battleship in the Philadelphia Experiment, moved forward with their own secret anti-gravity projects immediately after the war. It is interesting to note as well that Landig dismisses any flights to the Moon or Mars during the war or afterward, stating that the craft are for terrestrial flights. No mention is made of "making submarines fly." Can they fly to the Moon? Many researchers believe that they can.

Says Stevens on the ageing of the generals of the Third Power:

316   A V-1 rocket being perpared for launch in 1942

Dr. Hans Kammler was said to have died in 1972 according to Hans Rittermann. Otto Skorzeny died in 1975. Reinhard Gehlen died in 1979. Afterward, certainly, a younger generation of Nazis took over management of the Third Power. By the early-mid 1980's the Cold War was moving into its fourth decade with no end in sight or in contemplation in the minds of anyone on planet earth. Even the younger men running the Third Power must have been thinking about their own retirement as they were doubtless already in their 60s. It is no wonder they were more willing to divulge their secrets at this time. They were not sought criminals or marked men in any sort of way. They were free to fly first-class on commercial jets, they had no need of flying discs at this point.

The point is this postwar Nazi subculture existed in parallel alongside of both the Eastern and Western blocks of the Cold War, interacting and influencing both of them in ways which we are now only beginning to understand. Just as Farraday and Tesla postulated a self-inducting homopolar generator, these Nazis self-inducted the Cold War to a greater or lesser extent. And in some ways, especially in terms of technology, this postwar Nazi subculture was the focus of the conflict. And yet on the other hand, in terms of their middle position regarding communications, they were the glue which kept the Cold War from overheating. Was the Third Power was a balance-point, a fulcrum? Maybe this is overstating things but in the more narrow scope, especially in Europe and with high technology, it must have been important.

The Third Power was as much a child of the Cold War as were the Eastern and Western blocs and when one power got off this teeter-totter, shifting the world order, the existence of the Third Power as an active world-player became redundant. Now, twenty years later, their flying saucers rest in cold mountain storage, their U-boats rust in still-secret places, and their microfilm slowly deteriorates in hidden caches.

Today the Third Power is not trying to take over the world or even become a Fourth Reich. The Third Power was a concept for its members, and lives within friendly government officials and organizations of several countries as well as corporations and financial entities that owed a measure of their success to this Third Power. It also lives in the organizations who owe their roots or at least a greater part of their nourishment to Nazi ideas. This includes the Vril Society and the Karoteckia both of which, I am told by reliable sources, still exist. It also lives in the technology of those times, some of which must still be secret.[46]

The discussion of Nazi flying saucers resurfaced in August of 2018 when a series of declassified CIA files were reported by news outlets around the world including the *Daily Star* of the UK. The newspaper said on August 6, 2018 that the CIA files were about interviews with a German engineer named George Klein from between March 11 and May 20, 1952. Said the *Daily Star*:

In these interviews Klein claimed that the Nazis had a flying saucer that was capable of reaching heights of 12,400 meters in three minutes—with speeds of up to 2,500 mph. Nazi flying saucers appear twice in the CIA's trove of documents as part of their investigations into UFOs. The Klein testimony was published in newspapers in Greece, Iran and the Congo.

Klein said he was an engineer in the Ministry of Speer (i.e.: Albert Speer, Reich Minister of Armaments and War Production for Nazi Germany) and was present in Prague on February 14, 1945, at the first experimental flight of a flying saucer. Klein claims the Third Reich actually successfully carried out a test of their "flying saucer" in Prague on Valentine's Day, 1945, only months before the Czech capital was liberated by the Soviet Union's advancing Red Army.

Exceeding 2,500 mph (Mach 3) would make the Nazi saucer almost twice as fast as the state-of-the-art F-35 warplane being rolled out in the US and UK during World

War II. Klein claimed the saucer could takeoff vertically like a helicopter, and had been in development since 1941.

The CIA document alleges the saucers were constructed at the same slave-labor driven factories which made the dreaded V2 rockets. Nazi engineers were reportedly evacuated from Prague as the Red Army bore down upon them. Klein claims one team failed to be notified of the order to escape—and they were captured by the Soviets.

The CIA file reads: "Klein stated recently that though many people believe 'flying saucers' to be a postwar development, they were actually in the planning stage in German aircraft factories as early as 1941." "Klein was of the opinion that the 'saucers' are at present being constructed in accordance with German technical principles and expressed the belief that they will constitute serious competition to jet-propelled airplanes." Klein claimed the saucers were being built by the Russians, but no known Soviet saucers ever materialized.

Claims about the Nazis' advanced technology are often tied to alleged links between Hitler and the occult. Conspiracy theorists claim that remnants of the Third Reich fled to South America under the guidance of SS commander Hans Kammler. The Nazis are claimed to have continued testing their experimental weapons from secret bases in the Antarctic. UFO sightings from the era are alleged to be secret tests of experimental technology by the Nazis—and the US and Soviets. Theories persist to this day about advanced Reich technology—and still cause controversy.

Indeed, it would seem that this Third Power does still exist in South America, given the numerous UFO reports that come out of that mysterious continent. The craft can't all be from other planets. Rumors have existed for decades that the wealthy Tesla student Guglielmo Marconi had faked his death in 1937 and had moved with over a hundred European scientists to a secret jungle location in Venezuela or Colombia and set up a space center based on Tesla technology. This secret space base in South America was the plot of the 1979 James Bond film *Moonraker*.

And what now for the Third Power? Are the former German bases in Antarctica just empty bunkers waiting for intrepid explorers to find their dark caverns and light them up one more time? Were they destroyed by the Americans in the 1970s or earlier? We can assume that the secret bases in Tibet and the Arctic are abandoned as well. This may leave only the secret Haunebu bases in South America and perhaps the secret bases on King George Island and the Solomon Islands. Make no mistake, UFO activity in South America will continue for decades to come. Perhaps some of it will be extraterrestrials visiting the Earth or coming out of their deep underwater bases—if they exist. Others might be from the USA and the secret space program, while some South American governments, such as Chile, might have their own flying disk anti-gravity programs.

Still, other activity is bound to be Haunebu and Vril activity, plus the activity of the larger cigar-shaped craft. Yet, we have to wonder what this UFO activity signifies? We might surmise that the pilots are Volga Germans and the children of SS officers who escaped on submarines, ships and various aircraft to get to Argentina or Antarctica, but what are their missions? Is the Nazi psyops involving flying saucers not over yet as Joseph Farrell suggests in his books? Are they smuggling drugs and gold over international borders with these craft? Do they perform their own overflights over certain cities in South America to demonstrate their continued presence and power in these often poor countries?

Another question looms with regard to the issue of the secret Haunebu bases in Argentina, and even Haunebu factories near

A Haunebu on the ground from the Ettl documents.

Cordoba. It is clear that Juan Peron and a few other Argentinian generals must have known about these activities and allowed them. However, once Peron was out of power, most of this activity must have been kept completely secret, even from the Argentine government. Juan Peron was still active in politics in the 1970s, but with his death in 1974, much of the knowledge of the secret activities of the Germans in Argentina must have been lost. Otto Skorzeny died the next year in Spain and Rheinhard Gehlen died in 1979 (as has been noted, some have claimed that Skorzeny faked his death and lived in Florida for many more years).

Still, one must wonder what possible connections the Third Power has with the military of Argentina. Is it a game of cat and mouse? We see how the SS and the Third Reich had no beliefs in territorial integrity and were happy to put their secret facilities—many of them quite astonishing—in foreign countries and territories, even those of their adversaries. And, astonishing as many of the suppositions in this book have been, one of the most astonishing is that the Nazis, the SS, and the Thule/Vril Society had been working on these bases and technologies starting in the early 1930s with careful planning and lots financing, some of it from stolen gold and art.

As an extra-territorial power, they may feel that certain parts of Argentina, the Patagonian region mainly, are part of their territory and there is no doubt that large ranches and other patches of land in Paraguay, Chile, and even Bolivia and Brazil, are controlled by German families or companies. Paraguay as a country is notably pro-fascist and Hitler's birthday there is celebrated as Anti-Communist day.

Indeed, the remnants of the SS, the Black Sun community of super-mystics as described by Landig and partially by Miguel Serrano, must still exist in some form. A few dusty Haunebu craft must be sitting in a mountain hangar somewhere in the Andes, waiting to be fired up and flown into the sky. With the oil companies fading on the world stage and electric cars and trucks becoming the norm, maybe it is time for the technology behind the Haunebu, Vril and Andromeda craft to be brought to the public.

The main reason that this will not happen is that the secret space program, now called the Space Force, is using this technology and wants to keep it a military secret. For all the researchers into the secret German saucer programs and their aftermath the question has been: "Just how integrated into the American secret space program was the Nazi technology that we have chronicled in this book?"

With regard to Operation Paperclip, the Gehlen Org, Skorzeny, von Braun and others, it has been difficult to find the blurry edges between them and the CIA, Navy Intelligence, NASA, and the military space programs—now rolled into one as Space Force (NASA is not part of Space Force). There has always been talk of an early German base on our Moon. At this point, one would think that such a base would have become part of a network of secret American-Russian bases on the Moon.

One wants to think that the Black Sun Haunebu group and their Thule and Vril Society beliefs have tempered over the decades and they are not the cruel racist fascists that made WWII such a hideous war. The Japanese were equally cruel in the Pacific theater. Gehlen was part of the plots to assassinate Hitler, but he managed to escape Hitler's wrath when the plots failed. This helped him gain favor with the Americans when he surrendered to them at the end of the war. Along this vein, if Hitler had escaped

Reunion of the von Braun brothers with their father in Oberaudorf, Bavaria in February 1968. Left to right are Magnus, Baron von Braun, Sigismund and Werhner.

to Argentina and lived near Bariloche, it is unlikely that Gehlen would have ever visited him there. Gehlen spent most of his time in West Germany and rarely left the country. We are told that Bormann visited Hitler, and it is likely that Skorzeny did as well, but probably not Gehlen. As far as Hitler goes, if he were actually alive after the war, it seems he was unimportant and largely out of the loop with Gehlen, Skorzeny, Bormann, Udel and others as to the small fleet of U-boats, Haunebu, Vril and Andromeda craft.

As we go into the brave new world of tomorrow we can only hope that governments all over the world will come clean as to their involvement in the past history of UFO incidents since WWII, and any knowledge that they have of extraterrestrials as well. Perhaps they will acknowledge the secret bases on our Moon that they have had for decades—but that is another story.

A Vril craft photographed at the eruption of a volcano at Deception Island in Antarctica by an observation ship on December 4, 1967. This craft was probably coming from one of the secret bases in Antarctica such as King George Island.

324

FOOTNOTES & BIBLIOGRAPHY

1. *Donitz and the Wolf Packs*, Bernard Edwards, 1999, Hodder Headline, London.
2. *Hess and the Penguins,* Joseph P. Farrell, 2017, Adventures Unlimited Press, Kempton, IL.
3. *UFO Contact From Undersea*, Wendelle Stevens, 1982, UFO Photo Archives, Tucson, AZ.
4. *Grey Wolf: The Escape of Adolf Hitler*, Simon Dunstan and Gerrard Williams, 2011, Sterling Press, New York.
5. *Hitler's Terror Weapons: From VI to Vimana*, Geofrey Brooks, 2002, Pen and Sword Books, Barnsley, UK.
6. *Type VII U-boats*, Robert Stern, 1991, Arms and Armor Press, London.
7. *UFOs and Nukes*, Robert Hastings, 2017, (self-published on Amazon).
8. *German Secret Weapons of the Second World War,* Ian V. Hogg, 1999, Stackpole Books, Mechanicsburg, PA.
9. *The Ratline*, Peter Levenda, 2012, Ibis Press, Lakeworth, FL.
10. *Antarctica and the Secret Space Program*, David Childress, 2020. Adventures Unlimited Press, Kempton, IL.
11. *The U-Boat Wars*, Edwin P. Hoyt, 1984, Stein and Day Publishers, New York.
12. *Nazi International*, Joseph P. Farrell, 2008, Adventures Unlimited Press, Kempton, IL.
13. *The SS Totenkopf Ring,* Craig Gottlieb, 2008, Schiffer Publishing, Atglen, PA.
14. *Hidden Agenda,* Mike Bara, 2016, Adventures Unlimited Press, Kempton, IL.
15. *Vimana*, David Hatcher Childress, 2013, Adventures Unlimited Press, Kempton, IL.
16. *Man-Made UFOs,* Renato Vesco and David Childress, 1994, AUP, Kempton, IL.

17. *Flying Saucers: The Startling Evidence of the Invasion From Outer Space*. Coral Lorenzen, 1962,1966, Signet Books, New York.
18. *The Lubbock Lights,* David R. Wheeler, 1977, Award Books, New York.
19. *Hitler est vivo,* Ladislao Szabo, 1947, El Tábano, Buenos Aires. In Spanish.
20. *La Antárctica y otros Mitos* (The Antarctic and other myths), Miguel Serrano, 1948, Santiago (52 pagc booklet). In Spanish.
21. *UFO—Das Dritte Reich schlägt zurück?* (*UFO—The Third Reich Strikes Back?*) Norbert Jürgen Ratthofer and Ralf Ettl, 1989, Self-published. In German.
22. *Das Vril-Projekt. Der Endkampf um die Erde,* Norbert Jürgen Ratthofer and Ralf Ettl, 1992, Self-published. In German.
23. *Die Dunkle Seite Des Mondes* (*The Dark Side of the Moon*), Brad Harris, 1996, Pandora Books, Germany. In German.
24. *El Cordón Dorado: Hitlerismo Esotérico* (*The Golden Thread: Esoteric Hitlerism*), Miguel Serrano, 1978, Santiago. Part one of his Hitler Trilogy. In Spanish.
25. *Adolf Hitler, el Último Avatãra* (*Adolf Hitler: The Last Avatar*), Miguel Serrano, 1984, Santiago. Part two of his Hitler Trilogy. In Spanish.
26. *Manú: "Por El Hombre Que Vendra"* (*Manu: For the Coming Man*), Miguel Serrano, 1991, Santiago. Part three of his Hitler Trilogy. In Spanish.
27. *Memorias de Él y Yo vol. II, Adolf Hitler e la gran guerra* (*Memories of Him and Me. Volume 2. Adolf Hitler and the Great War*), Miguel Serrano, 1997, Santiago. In Spanish.
28. *Black Sun: Aryan Cults, Esoteric Nazism, and the Politics of Identity*, Nicholas Goodrick-Clarke, 2002, New York University Press, New York.
29. *The Morning of the Magicians*, Jacques Bergier and Louis Pauwels, 1960, 1963, English edition, Stein and Day, New York.
30. *Reich of the Black Sun,* Joseph P. Farrell, 2009, AUP, Kempton, IL.
31. *Casebook on the Men in Black*, Jim Keith, 1997, AUP, Kempton, IL.
32. *Remarkable Luminous Phenomena in Nature*, William Corliss, 2001, Sourcebook Project, Glen Arm, MD.

33. *Hitler: The Survival Myth*, Donald M. McKale, 1983, Stein & Day, New York.
34. *Hitler's Flying Saucers*, Henry Stevens, 2003, Adventures Unlimited Press, Kempton, IL.
35. *Arktos: The Polar Myth*, Joscelyn Godwin, 1996, AUP, Kempton, IL.
36. *Electric UFOs*, Albert Budden, 1998, Blandford Books, London.
37. *Himmler's Crusade: The Nazi Expedition to Find the Origins of the Aryan Race*, Christopher Hale, 2003, John Wiley & Sons, Hoboken, NJ.
38. *Saucers, Swastikas and Psyops*, Joseph P. Farrell, 2011, Adventures Unlimited Press, Kempton, IL.
39. *Roswell and the Reich,* Joseph P. Farrell, 2010, AUP, Kempton, IL.
40. *The CIA UFO Papers,* Dan Wright, 2019, MUFON-Red Wheel-Weiser, Newburyport, MA.
41. *Underground Bases and Tunnels*, Richard Sauder, 1997, AUP, Kempton, IL.
42. *Underwater and Underground Bases*, Richard Sauder, 2001, AUP, Kempton, IL.
43. *The Great Heroin Coup: Drugs, Intelligence and International Fascism,* Henrik Kruger, 1980, South End Press, Boston.
44. *Dark Fleet*, Len Kasten, 2020, Bear & Company, Rochester, VT.
45. *Vimana: Flying Machines of the Ancients*, David Hatcher Childress, 2013, AUP, Kempton, IL.
46. *Dark Star*, Henry Stevens, 2011, Adventures Unlimited Press, Kempton, IL.
48. *The German Saucer Story*, Michael Barton, 1968, Future Press, Los Angeles.
49. *We Want You: Is Hitler Still Alive?,* Michael Barton, 1960, Future Press, Los Angeles.
50. *Emerald Cup—Ark of Gold*, Col. Howard Buechner, 1991, Thunderbird Press, Metairie, LA.
51. *Il Deutsche Flugscheiben und U-Boote Ueberwachen Die Weltmeere,* O. Bergmann, 1989, Hugin Publishing, Germany.

52. *Anti-Gravity & the World Grid*, David Hatcher Childress, 1987, AUP, Kempton, IL.
53. *Vril, The Power of the Coming Race*, Sir Edward Bulwer-Lytton, 1871, Blackwood and Sons, London.
54. *The Hunt for Zero Point*, Nick Cook, 2003, Broadway Books, New York.
55. *Secret Cities of Old South America*, Harold Wilkins, 1952, London, reprinted 1998, Adventures Unlimited Press, Kempton, IL.
56. *War in Ancient India*, V. R. Dikshitar, 1944, Oxford University Press (1987 edition published by Motilal Banarsidass, Delhi).
57. *Invisible Residents*, Ivan T. Sanderson, 1970, Adventures Unlimited Press, Kempton, IL.
58. *The Philadelphia Experiment*, William Moore and Charles Berlitz, 1979, Grosset & Dunlap, New York.
59. *UFO Photographs Around the World*, Edited by Wendelle Stevens, 1986, UFO Photo Archives, Tucson, AZ.
60. *UFO Photographs Around the World, Vol. II*, Edited by Wendelle Stevens, 1986, UFO Photo Archives, Tucson, AZ.
61. *UFO Photographs Around the World, Vol. III,* Edited by Wendelle Stevens, 1986, UFO Photo Archives, Tucson, AZ.

A plastic model kit for the Haunebu II saucer.

REPRODUCED AT THE NATIONAL ARCHIVES

DECLASSIFIED
Authority: NND 813 055
By K.C. NARA Date 2/4/97

2-5317.

TOP SECRET

| USAFE 14 | TT 1524 | TOP SECRET | 4 Nov 1948 |

From OI OB

For some time we have been concerned by the recurring reports on flying saucers. They periodically continue to crop up; during the last week, one was observed hovering over Neubiberg Air Base for about thirty minutes. They have been reported by so many sources and from such a variety of places that we are convinced that they cannot be disregarded and must be explained on some basis which is perhaps slightly beyond the scope of our present intelligence thinking.

When officers of this Directorate recently visited the Swedish Air Intelligence Service. This question was put to the Swedes. Their answer was that some reliable and fully technically qualified people have reached the conclusion that "these phenomena are obviously the result of a high technical skill which cannot be credited to any presently known culture on earth." They are therefore assuming that these objects originate from some previously unknown or unidentified technology, possibly outside the earth.

One of these objects was observed by a Swedish technical expert near his home on the edge of a lake. The object crashed or landed in the lake and he carefully noted its azimuth from his point of observation. Swedish intelligence was sufficiently confident in his observation that a naval salvage team was sent to the lake. Operations were underway during the visit of USAFE officers. Divers had discovered a previousuly uncharted crater on the floor of the lake. No further information is available, but we have been promised knowledge of the results. In their opinion, the observation was reliable, and they believe that the depression on the floor of the lake, which did not appear on current Hydrographic charts, was in fact caused by a flying saucer.

Although accepting this theory of the origin of these objects poses a whole new group of questions and puts much of our thinking in a changed light, we are inclined not to discredit entirely this somewhat spectacular theory, meantime keeping an open mind on the subject. What are your reactions?

TOP SECRET

(END OF USAFE ITEM 14)

A declassified Top Secret Air Force document dated November 4, 1948 discussing flying saucers and whether there was any power on Earth that had this technology.

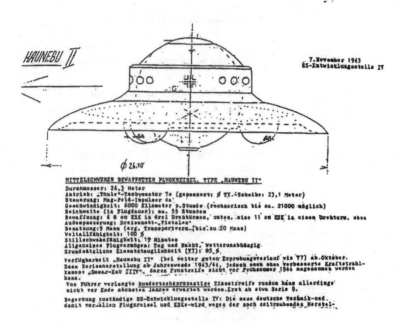

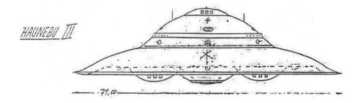

ANTARCTICA AND THE SECRET SPACE PROGRAM
By David Hatcher Childress

David Childress, popular author and star of the History Channel's show *Ancient Aliens*, brings us the incredible tale of Nazi submarines and secret weapons in Antarctica and elsewhere. He then examines Operation High-Jump with Admiral Richard Byrd in 1947 and the battle that he apparently had in Antarctica with flying saucers. Through "Operation Paperclip," the Nazis infiltrated aerospace companies, banking, media, and the US government, including NASA and the CIA after WWII. Does the US Navy have a secret space program that includes huge ships and hundreds of astronauts?

392 Pages. 6x9 Paperback. Illustrated. $22.00 Code: ASSP

NORTH CAUCASUS DOLMENS
By Boris Loza, Ph.D.

Join Boris Loza as he travels to his ancestral homeland to uncover and explore dolmens firsthand. Chapters include: Ancient Mystic Megaliths; Who Built the Dolmens?; Why the Dolmens were Built; Asian Connection; Indian Connection; Greek Connection; Olmec and Maya Connection; Sun Worshippers; Dolmens and Archeoastronomy; Location of Dolmen Quarries; Hidden Power of Dolmens; and much more! Tons of Illustrations! A fascinating book of little-seen megaliths. Color section.

252 Pages. 5x9 Paperback. Illustrated. $24.00. Code NCD

THE ENCYCLOPEDIA OF MOON MYSTERIES
Secrets, Anomalies, Extraterrestrials and More
By Constance Victoria Briggs

Our moon is an enigma. The ancients viewed it as a light to guide them in the darkness, and a god to be worshipped. Did you know that: Aristotle and Plato wrote about a time when there was no Moon? Several of the NASA astronauts reported seeing UFOs while traveling to the Moon?; the Moon might be hollow?; Apollo 10 astronauts heard strange "space music" when traveling on the far side of the Moon?; strange and unexplained lights have been seen on the Moon for centuries?; there are said to be ruins of structures on the Moon?; there is an ancient tale that suggests that the first human was created on the Moon?; Tons more. Tons of illustrations with A to Z sections for easy reference and reading.

152 Pages. 7x10 Paperback. Illustrated. $19.95. Code: EOMM

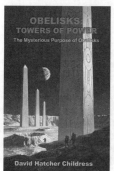

OBELISKS: TOWERS OF POWER
The Mysterious Purpose of Obelisks
By David Hatcher Childress

Some obelisks weigh over 500 tons and are massive blocks of polished granite that would be extremely difficult to quarry and erect even with modern equipment. Why did ancient civilizations in Egypt, Ethiopia and elsewhere undertake the massive enterprise it would have been to erect a single obelisk, much less dozens of them? Were they energy towers that could receive or transmit energy? With discussions on Tesla's wireless power, and the use of obelisks as gigantic acupuncture needles for earth, Chapters include: Megaliths Around the World and their Purpose; The Crystal Towers of Egypt; The Obelisks of Ethiopia; Obelisks in Europe and Asia; Mysterious Obelisks in the Americas; The Terrible Crystal Towers of Atlantis; Tesla's Wireless Power Distribution System; Obelisks on the Moon; more. 8-page color section.

336 Pages. 6x9 Paperback. Illustrated. $22.00 Code: OBK

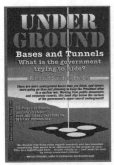

UNDERGROUND BASES & TUNNELS:
What is the Government Trying to Hide?
by Richard Sauder, Ph.D.

Working from government documents and corporate records, Sauder has compiled an impressive book that digs below the surface of the military's super-secret underground! Go behind the scenes into little-known corners of the public record and discover how corporate America has worked hand-in-glove with the Pentagon for decades, dreaming about, planning, and actually constructing, secret underground bases. This book includes chapters on the locations of the bases, the tunneling technology, various military designs for underground bases, abductions, needles & implants, military involvement in "alien" cattle mutilations, more. 50-page photo & map insert.

201 pages. 6x9 Paperback. Illustrated. $15.95. Code: UGB

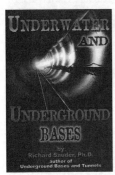

UNDERWATER & UNDERGROUND BASES
by Richard Sauder, Ph.D.

Dr. Sauder lays out the amazing evidence and government paper trail for the construction of huge, manned bases offshore, in mid-ocean, and deep beneath the sea floor! Official United States Navy documents, and other hard evidence, raise many questions about what really lies 20,000 leagues beneath the sea. Plus, breakthrough material reveals the existence of additional clandestine underground facilities as well as the surprising location of one of the CIA's own underground bases. Plus, information on tunneling and cutting-edge, high speed rail magnetic-levitation (MagLev) technology.

264 pages. 6x9 Paperback. Illustrated. $16.95. Code: UUB

AMERICAN CONSPIRACY FILES
The Stories We Were Never Told
By Peter Kross

Kross reports on conspiracies in the Revolutionary War, including those surrounding Benedict Arnold and Ben Franklin's son, William. He delves into the large conspiracy to kill President Lincoln and moves into our modern day with chapters on the deaths of JFK, RFK and MLK., the reasons behind the Oklahoma City bombing, the sordid plots of President Lyndon Johnson and more. Chapters on Edward Snowden; The Weather Underground; Patty Hearst; The Death of Mary Meyer; Marilyn Monroe; The Zimmerman Telegram; BCCI; Operation Northwinds; The Search for Nazi Gold; The Death of Frank Olsen; tons more. Over 50 chapters in all.

460 Pages. 6x9 Paperback. Illustrated. $19.95 Code: ACF

DARK STAR
By Henry Stevens

WWII expert Stevens takes us through the final twists and turns of the fall of the Third Reich and the fate of its secret bases, technology and vast U-Boat fleet. Stevens reveals secret submarine bases in Greenland and the Canary Islands. He looks at the escape of Nazis to South America, then turns to the Haunebu flying disk. He discusses Hans Kammler and the last battalion with its saucer technology—derived from the work of Nikola Tesla—and the secret submarine fleet that continued after the official end of the war. Chapters include: German Secret Bases; Norway; Greenland; The Last Battalion; Fuerteventura; Electric Schauberger; South America and Otto Skorzeny; The Third Power in Practice; tons more.

342 Pages. 6x9 Paperback. Illustrated. $19.95 Code: DSTR

HESS AND THE PENGUINS
The Holocaust, Antarctica and the Strange Case of Rudolf Hess
By Joseph P. Farrell

Farrell looks at Hess' mission to make peace with Britain and get rid of Hitler—even a plot to fly Hitler to Britain for capture! How much did Göring and Hitler know of Rudolf Hess' subversive plot, and what happened to Hess? Why was a doppleganger put in Spandau Prison and then "suicided"? Did the British use an early form of mind control on Hess' double? John Foster Dulles of the OSS and CIA suspected as much. Farrell also uncovers the strange death of Admiral Richard Byrd's son in 1988, about the same time of the death of Hess.

288 Pages. 6x9 Paperback. Illustrated. $19.95. Code: HAPG

HIDDEN FINANCE, ROGUE NETWORKS & SECRET SORCERY
The Fascist International, 9/11, & Penetrated Operations
By Joseph P. Farrell

Farrell investigates the theory that there were not *two* levels to the 9/11 event, but *three*. He says that the twin towers were downed by the force of an exotic energy weapon, one similar to the Tesla energy weapon suggested by Dr. Judy Wood, and ties together the tangled web of missing money, secret technology and involvement of portions of the Saudi royal family. Farrell unravels the many layers behind the 9-11 attack, layers that include the Deutschebank, the Bush family, the German industrialist Carl Duisberg, Saudi Arabian princes and the energy weapons developed by Tesla before WWII.

296 Pages. 6x9 Paperback. Illustrated. $19.95. Code: HFRN

THRICE GREAT HERMETICA & THE JANUS AGE
By Joseph P. Farrell

What do the Fourth Crusade, the exploration of the New World, secret excavations of the Holy Land, and the pontificate of Innocent the Third all have in common? Answer: Venice and the Templars. What do they have in common with Jesus, Gottfried Leibniz, Sir Isaac Newton, Rene Descartes, and the Earl of Oxford? Answer: Egypt and a body of doctrine known as Hermeticism. The hidden role of Venice and Hermeticism reached far and wide, into the plays of Shakespeare (a.k.a. Edward DeVere, Earl of Oxford), into the quest of the three great mathematicians of the Early Enlightenment for a lost form of analysis, and back into the end of the classical era, to little known Egyptian influences at work during the time of Jesus.

354 Pages. 6x9 Paperback. Illustrated. $19.95. Code: TGHJ

HITLER'S SUPPRESSED AND STILL-SECRET WEAPONS, SCIENCE AND TECHNOLOGY
by Henry Stevens

In the closing months of WWII the Allies assembled mind-blowing intelligence reports of supermetals, electric guns, and ray weapons able to stop the engines of Allied aircraft—in addition to feared x-ray and laser weaponry. Chapters include: The Kammler Group; German Flying Disc Update; The Electromagnetic Vampire; Liquid Air; Synthetic Blood; German Free Energy Research; German Atomic Tests; The Fuel-Air Bomb; Supermetals; Red Mercury; Means to Stop Engines; more.

335 Pages. 6x9 Paperback. Illustrated. $19.95. Code: HSSW

ROSWELL AND THE REICH
The Nazi Connection
By Joseph P. Farrell

Farrell has meticulously reviewed the best-known Roswell research from UFO-ET advocates and skeptics alike, as well as some little-known source material, and comes to a radically different scenario of what happened in Roswell, New Mexico in July 1947, and why the US military has continued to cover it up to this day. Farrell presents a fascinating case sure to disturb both ET believers and disbelievers, namely, that what crashed may have been representative of an independent postwar Nazi power—an extraterritorial Reich monitoring its old enemy, America, and the continuing development of the very technologies confiscated from Germany at the end of the War.

540 pages. 6x9 Paperback. Illustrated. $19.95. Code: RWR

HIDDEN AGENDA
NASA and the Secret Space Program
By Mike Bara

Bara delves into secret bases on the Moon, and exploring the many other rumors surrounding the military's secret projects in space. On June 8, 1959, a group at the ABMA produced for the US Department of the Army a report entitled Project Horizon, a "Study for the Establishment of a Lunar Military Outpost." The permanent outpost was predicted to cost $6 billion and was to become operational in December 1966 with twelve soldiers stationed at the Moon base. Does hacker Gary Mackinnon's discovery of defense department documents identifying "non-terrestrial officers" serving in space? Includes an 8-page color section.

346 Pages. 6x9 Paperback. Illustrated. $19.95. Code: HDAG

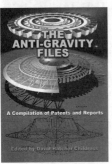

THE ANTI-GRAVITY FILES
A Compilation of Patents and Reports
Edited by David Hatcher Childress

In the tradition of *The Anti-Gravity Handbook* and *the Time-Travel Handbook* comes this compilation of material on anti-gravity, free energy, flying saucers and Tesla technology. With plenty of technical drawings and explanations, this book reveals suppressed technology that will change the world in ways we can only dream of. Chapters include: A Brief History of Anti-Gravity Patents; The Motionless Electromagnet Generator Patent; Mercury Anti-Gravity Gyros; The Tesla Pyramid Engine; Anti-Gravity Propulsion Dynamics; The Machines in Flight; More Anti-Gravity Patents; Death Rays Anyone?; The Unified Field Theory of Gravity; and tons more. Heavily illustrated. 4-page color section.

216 pages. 8x10 Paperback. Illustrated. $22.00. Code: AGF

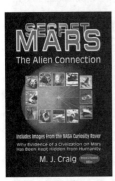

SECRET MARS: The Alien Connection
By M. J. Craig

While scientists spend billions of dollars confirming that microbes live in the Martian soil, people sitting at home on their computers studying the Mars images are making far more astounding discoveries... they have found the possible archaeological remains of an extraterrestrial civilization. Hard to believe? Well, this challenging book invites you to take a look at the astounding pictures yourself and make up your own mind. *Secret Mars* presents over 160 incredible images taken by American and European spacecraft that reveal possible evidence of a civilization that once lived, and may still live, on the planet Mars... powerful evidence that scientists are ignoring! A visual and fascinating book!

352 Pages. 6x9 Paperback. Illustrated. $19.95. Code: SMAR

ANCIENT ALIENS ON THE MOON
By Mike Bara
What did NASA find in their explorations of the solar system that they may have kept from the general public? How ancient really are these ruins on the Moon? Using official NASA and Russian photos of the Moon, Bara looks at vast cityscapes and domes in the Sinus Medii region as well as glass domes in the Crisium region. Bara also takes a detailed look at the mission of Apollo 17 and the case that this was a salvage mission, primarily concerned with investigating an opening into a massive hexagonal ruin near the landing site. Chapters include: The History of Lunar Anomalies; The Early 20th Century; Sinus Medii; To the Moon Alice!; Mare Crisium; Yes, Virginia, We Really Went to the Moon; Apollo 17; more. Tons of photos of the Moon examined for possible structures and other anomalies.
248 Pages. 6x9 Paperback. Illustrated.. $19.95. Code: AAOM

ANCIENT ALIENS ON MARS
By Mike Bara
Bara brings us this lavishly illustrated volume on alien structures on Mars. Was there once a vast, technologically advanced civilization on Mars, and did it leave evidence of its existence behind for humans to find eons later? Did these advanced extraterrestrial visitors vanish in a solar system wide cataclysm of their own making, only to make their way to Earth and start anew? Was Mars once as lush and green as the Earth, and teeming with life? Chapters include: War of the Worlds; The Mars Tidal Model; The Death of Mars; Cydonia and the Face on Mars; The Monuments of Mars; The Search for Life on Mars; The True Colors of Mars and The Pathfinder Sphinx; more. Color section.
252 Pages. 6x9 Paperback. Illustrated. $19.95. Code: AMAR

ANCIENT ALIENS ON MARS II
By Mike Bara
Using data acquired from sophisticated new scientific instruments like the Mars Odyssey THEMIS infrared imager, Bara shows that the region of Cydonia overlays a vast underground city full of enormous structures and devices that may still be operating. He peels back the layers of mystery to show images of tunnel systems, temples and ruins, and exposes the sophisticated NASA conspiracy designed to hide them. Bara also tackles the enigma of Mars' hollowed out moon Phobos, and exposes evidence that it is artificial. Long-held myths about Mars, including claims that it is protected by a sophisticated UFO defense system, are examined. Data from the Mars rovers Spirit, Opportunity and Curiosity are examined; everything from fossilized plants to mechanical debris is exposed in images taken directly from NASA's own archives.
294 Pages. 6x9 Paperback. Illustrated. $19.95. Code: AAM2

ANCIENT TECHNOLOGY IN PERU & BOLIVIA
By David Hatcher Childress
Childress speculates on the existence of a sunken city in Lake Titicaca and reveals new evidence that the Sumerians may have arrived in South America 4,000 years ago. He demonstrates that the use of "keystone cuts" with metal clamps poured into them to secure megalithic construction was an advanced technology used all over the world, from the Andes to Egypt, Greece and Southeast Asia. He maintains that only power tools could have made the intricate articulation and drill holes found in extremely hard granite and basalt blocks in Bolivia and Peru, and that the megalith builders had to have had advanced methods for moving and stacking gigantic blocks of stone, some weighing over 100 tons.
340 Pages. 6x9 Paperback. Illustrated.. $19.95 Code: ATP

ORDER FORM

**10% Discount
When You Order
3 or More Items!**

One Adventure Place
P.O. Box 74
Kempton, Illinois 60946
United States of America
Tel.: 815-253-6390 • Fax: 815-253-6300
Email: auphq@frontiernet.net
http://www.adventuresunlimitedpress.com

ORDERING INSTRUCTIONS

✓ Remit by USD$ Check, Money Order or Credit Card

✓ Visa, Master Card, Discover & AmEx Accepted

✓ Paypal Payments Can Be Made To:

   info@wexclub.com

✓ Prices May Change Without Notice

✓ 10% Discount for 3 or More Items

SHIPPING CHARGES

United States

✓ POSTAL BOOK RATE

✓ Postal Book Rate { $4.50 First Item / 50¢ Each Additional Item

✓ Priority Mail { $7.00 First Item / $2.00 Each Additional Item

✓ UPS { $9.00 First Item (Minimum 5 Books) / $1.50 Each Additional Item

   NOTE: UPS Delivery Available to Mainland USA Only

Canada

✓ Postal Air Mail { $19.00 First Item / $3.00 Each Additional Item

✓ Personal Checks or Bank Drafts MUST BE

   US$ and Drawn on a US Bank

✓ Canadian Postal Money Orders OK

✓ Payment MUST BE US$

All Other Countries

✓ Sorry, No Surface Delivery!

✓ Postal Air Mail { $19.00 First Item / $7.00 Each Additional Item

✓ Checks and Money Orders MUST BE US$ and Drawn on a US Bank or branch.

✓ Paypal Payments Can Be Made in US$ To:

   info@wexclub.com

SPECIAL NOTES

✓ RETAILERS: Standard Discounts Available

✓ BACKORDERS: We Backorder all Out-of-Stock Items Unless Otherwise Requested

✓ PRO FORMA INVOICES: Available on Request

✓ DVD Return Policy: Replace defective DVDs only

ORDER ONLINE AT: www.adventuresunlimitedpress.com

**10% Discount When You Order
3 or More Items!**

Please check: ✓

☐ This is my first order   ☐ I have ordered before

Name			
Address			
City			
State/Province		Postal Code	
Country			
Phone: Day		Evening	
Fax	Email		

Item Code	Item Description	Qty	Total

	Subtotal ▶	
Please check: ✓	Less Discount-10% for 3 or more items ▶	
☐ Postal-Surface	Balance ▶	
☐ Postal-Air Mail Illinois Residents 6.25% Sales Tax ▶		
(Priority in USA) Previous Credit ▶		
☐ UPS Shipping ▶		
(Mainland USA only) Total (check/MO in USD$ only) ▶		

☐ Visa/MasterCard/Discover/American Express

Card Number:

Expiration Date: Security Code:

‾ ‾ ‾ ‾ ‾ ‾ ‾ ‾ ‾ ‾ ‾ ‾ ‾ ‾ ‾ ‾ ‾ ‾
✓ SEND A CATALOG TO A FRIEND: